SpringerBriefs in Earth System Sciences

Series editors

Jorge Rabassa, Ushuaia, Argentina
Gerrit Lohmann, Bremen, Germany
Justus Notholt, Bremen, Germany
Lawrence A. Mysak, Montreal, Canada
Vikram Unnithan, Bremen, Germany

More information about this series at http://www.springer.com/series/10032

Silvia Leonor Lagorio · Haroldo Vizán
Silvana Evangelina Geuna

Early Cretaceous Volcanism in Central and Eastern Argentina During Gondwana Break-Up

 Springer

Silvia Leonor Lagorio
Instituto de Geología y Recursos Minerales
Servicio Geológico Minero Argentino
    (IGRM-SEGEMAR), Parque Tecnológico
    Miguelete
San Martín, Buenos Aires
Argentina

Silvana Evangelina Geuna
IGEBA (CONICET-UBA)
Departamento de Ciencias Geológicas,
    Facultad de Ciencias Exactas y Naturales,
    Universidad de Buenos Aires
Ciudad Autónoma de Buenos Aires
Argentina

Haroldo Vizán
IGEBA (CONICET-UBA)
Departamento de Ciencias Geológicas,
    Facultad de Ciencias Exactas y Naturales,
    Universidad de Buenos Aires
Ciudad Autónoma de Buenos Aires
Argentina

ISSN 2191-589X   ISSN 2191-5903   (electronic)
SpringerBriefs in Earth System Sciences
ISBN 978-3-319-29591-6   ISBN 978-3-319-29593-0   (eBook)
DOI 10.1007/978-3-319-29593-0

Library of Congress Control Number: 2016930287

Printed on acid-free paper

This Springer imprint is published by SpringerNature
The registered company is Springer International Publishing AG Switzerland

Preface

The objective of this contribution is to analyse the geochemistry, petrogenesis and geodynamics of the Early Cretaceous volcanism in Argentina, mainly located in Córdoba and Misiones provinces, during the break-up of Gondwana.

In Córdoba (central Argentina), the analysed volcanic rocks are outcropping in Sierra Chica and different groups of lithological types were recognized in various localities. These are mainly alkaline basalts that reflect a lithospheric mantle source.

In Misiones (north-eastern Argentina), the volcanic rocks are tholeiitic basalts and belong to Paraná Magmatic Province (PMP) which is regionally extended in South America and has a counterpart in Africa (in the localities of Etendeka and Angola). This large igneous province (LIP) has been widely studied by several authors, and different models have been considered to explain its origin. Dating obtained through various methodologies ($^{40}Ar/^{39}Ar$, U–Pb, and Re–Os isochrons) and published by different authors indicate an Early Cretaceous age between 131.6 ± 2.3 and 134.7 ± 1 Ma for PMP, though interbedded acid volcanic rocks yielded even 137.3 ± 1.8 Ma.

A new $^{40}Ar/^{39}Ar$ age of 129.6 ± 1 Ma from an alkaline rock of Sierra Chica of Córdoba (SCC) presented in this contribution, points out that this volcanism was slightly younger than PMP.

It is suggested that the volcanism in Misiones and in the overall LIP could have been linked to an ascending limb of a large-scale convective roll induced by the subduction in the western margin of Gondwana. This ascending limb might have mainly affected weak cortical areas (old sutures between cratons). On the other hand, the volcanism of Sierra Chica might have been related to a small-scale edge-driven convection triggered by the great contrast in thickness between the Río de la Plata craton and the Pampia terrane.

Acknowledgments

This monograph is dedicated to the memory of Enzo Piccirillo, Carlos Ernesto Gordillo and Daniel Alberto Valencio, from whom the authors learnt how to work in science. Enzo Piccirillo and Marco Iacumin are specially acknowledged for the helpful discussions and reviews of the Ph.D. thesis of the first author, as well as for the support in technical, analytical and field work received from the University of Trieste. Special gratitude for Giuliano Bellieni for microprobe analyses carried out at the University of Padova. In the same way, the following are kindly acknowledged: Marcela Remesal, Sonia Quenardelle, Stella Poma, Víctor Ramos, Carlos Rapela, José Viramonte, Mónica Escayola and Iván Petrinovic for significant opinions and suggestions expressed in diverse opportunities. We also thank SEGEMAR, CONICET and Universidad de Buenos Aires for letting us do this work, particularly to Alberto Ardolino and Jose Mendía, for the encouragement received to undertake the study of Misiones rocks. Kindly thanks to Claudio Gaucher and Jorge Bossi for the significant comments on the Precambrian units in south-eastern South America. This work was supported by grants from Universidad de Buenos Aires (UBACyT 20020100100894) and Consejo Nacional de Investigaciones Científicas y Técnicas (CONICET-PIP 112-200801-02828).

Contents

Chapter 1
Introduction

Abstract During the Early Cretaceous, a reorganization of lithospheric plates led to one of the major volcanic processes: the Paraná Magmatic Province (PMP) that covered an extensive region of Brazil, Paraguay, Uruguay and north-eastern Argentina. It was a tholeiitic event constituting a large igneous province (LIP). By contrast, an alkaline volcanism also occurred but it is volumetrically restricted, has a peripheral location with respect to this LIP and took place either prior as well as contemporary and posthumously to that large tholeiitic episode. Both volcanic events are Early Cretaceous and must be linked to the break-up of Western Gondwana. In the Province of Córdoba (central Argentina), alkaline volcanic rocks outcrop in Sierra Chica (SCC), and were generated under extensional conditions integrating a system of rifts. They are located about 150 km west of PMP basalts that are lying in the sub-surface of Chaco-Paraná basin. Recent radiometric dating provided by other authors revealed ages of 134.7 ± 1 and 131.6 ± 2.3 Ma for PMP basalts, whereas interbedded acid rocks show 137.3 ± 1.8 and 134.3 ± 0.8 Ma, respectively. A new $^{40}Ar/^{39}Ar$ dating of 129 ± 1 Ma is here presented for the volcanism of SCC, indicating that it is slightly younger than that of the PMP. Both volcanic events must have had different origins in the context of reorganization of geological plates during the Early Cretaceous.

Keywords Tholeiitic basalts · Alkaline basalts · Cretaceous · Paraná · Córdoba · Argentina

1.1 Introduction

During the Cretaceous, there was a reorganization of lithospheric plates that led to one of the major volcanic processes in our planet. The volcanism of the Paraná basin in South America covered an extensive region of Brazil, Paraguay, Uruguay and north-eastern Argentina. It was a major tholeiitic event, therefore constitutes a large igneous province (LIP) called Paraná–Etendeka (Bellieni et al. 1984; Erlank et al. 1984), Paraná–Etendeka–Angola (PEA, Marzoli et al. 1999) and Paraná Magmatic Province, only for South America (PMP, Marques et al. 1999).

© The Author(s) 2016
S.L. Lagorio et al., *Early Cretaceous Volcanism in Central and Eastern Argentina During Gondwana Break-Up*, SpringerBriefs in Earth System Sciences, DOI 10.1007/978-3-319-29593-0_1

By contrast, alkaline volcanism is volumetrically restricted, has a peripheral location with respect to the volcanism of Paraná basin and has developed prior, contemporary as well as posthumously to the main tholeiitic event (Fig. 1.1a).

Both, the tholeiitic and the alkaline Early Cretaceous volcanic events are linked to the break-up of Western Gondwana (South America and Africa), which finally led to the opening of the South Atlantic Ocean (see Piccirillo and Melfi 1988, among others).

In the Province of Córdoba (central Argentina), alkaline volcanic rocks outcrop in Sierra Chica, where a sedimentary–volcanic complex is exposed at about 150 km west of the Chaco-Paraná basin, which is the southern portion of the entire Paraná basin (Fig. 1.1a, b).

Consequently, the alkaline volcanism of Córdoba was developed under extensional conditions (Uliana et al. 1990), integrating the so-called Central System of Rifts (Rossello and Mozetic 1999; Ramos 1999; Fig. 1.1b).

Other Lower Cretaceous basalts that belong to this Central System are buried in the southern part of Córdoba Province, within the General Levalle basin (Fig. 1.1b; Webster et al. 2004; Chebli et al. 2005). These volcanic rocks were recognized throughout a well that was drilled to explore the petroleum potential of that basin. There are no geochemical studies on these rocks, but their petrographic characteristics closely resemble those outcropping in Sierra Chica.

West of Córdoba Province, in Sierra de las Quijadas and Cerrillada de las Cabras (San Luis Province; Fig. 1.1a), other Lower Cretaceous basaltic rocks are outcropping. They appear as lava flows and dykes interbedded with a sedimentary sequence within Las Salinas depocenter, belonging to the Western System of Rifts (Fig. 1.1b), and are comparable to those outcropping in the Sierra Chica of Córdoba (Gordillo 1972; Costa et al. 2001), also from their geochemical features (Martínez et al. 2012).

Recent $^{40}Ar/^{39}Ar$ dating reveals an age of 134.7 ± 1 Ma for rocks of PMP and a probably duration of the volcanic event involved in this LIP of 1.2 million years (Thiede and Vasconcelos 2010). The dating is consistent with other previous ones, mainly between 133 and 130 Ma (e.g. Renne et al. 1992, 1996; Ernesto et al. 1999), after applying the method of recalculation proposed by Kuiper et al. (2008). On the other hand, U-Pb dating obtained in the acid volcanic rocks interbedded with the tholeiitic basalts varies between 137.3 ± 1.8 and 134.3 ± 0.8 Ma (Wildner et al. 2006; Janasi et al. 2011), whereas a Re–Os isochron presented by Rocha-Junior et al. (2012) indicates a younger age of 131.6 ± 2.3 Ma.

Based on whole rock, K/Ar dating of the lavas of Sierra Chica points out that they have been extruded mainly from 136 to 115 Ma (e.g. Gordillo and Lencinas 1967a, b, 1969; Linares and Valencio 1974; Cejudo Ruiz et al. 2006). In this contribution, a new $^{40}Ar/^{39}Ar$ dating of 129.6 ± 1 Ma is presented, obtained in sanidine phenocrysts of a sample taken at the locality of Almafuerte, that exposes that the volcanism in Córdoba Province is slightly younger than the great magmatic event of PMP.

Therefore, both eruptive processes occurred mainly during the Early Cretaceous, relatively close in age, during the global extensional tectonics that caused the fragmentation of Pangea, but possibly through different geodynamic mechanisms.

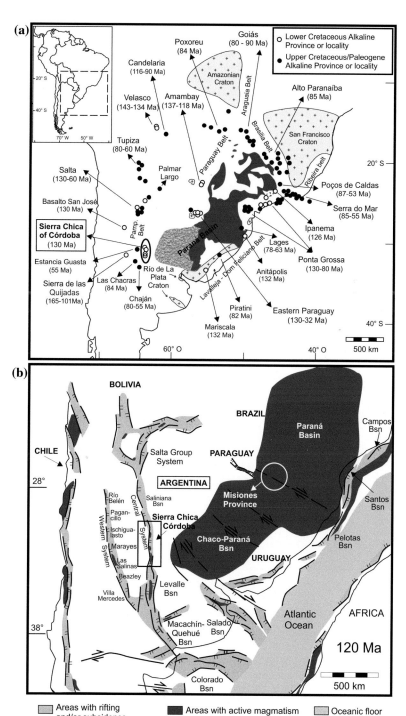

(a)

Lower Cretaceous Alkaline Province or locality
Upper Cretaceous/Paleogene Alkaline Province or locality

Poxoreu (84 Ma)
Goiás (80 - 90 Ma)
Candelaria (116-90 Ma)
Amazonian Craton
Alto Paranaíba (85 Ma)
Velasco (143-134 Ma)
Amambay (137-118 Ma)
Araguaia Belt
Brasília Belt
San Francisco Craton
Tupiza (80-60 Ma)
Paraguay Belt
Ribeira belt
20° S
Salta (130-60 Ma)
Palmar Largo
Poços de Caldas (87-53 Ma)
Basalto San José (130 Ma)
Serra do Mar (85-55 Ma)
Sierra Chica of Córdoba (130 Ma)
Pamp. Belt
Paraná Basin
Ipanema (126 Ma)
Estancia Guasta (55 Ma)
Río de La Plata Craton
Lavalleja - Dom Feliciano Belt
Lages (78-63 Ma)
Ponta Grossa (130-80 Ma)
Sierra de las Quijadas (165-101Ma)
Las Chacras (84 Ma)
Anitápolis (132 Ma)
Chaján (80-55 Ma)
Piratini (82 Ma)
Eastern Paraguay (130-32 Ma)
Mariscala (132 Ma)
40° S
60° O
40° O
500 km

20° S
40° S
70° W   50° W

(b)

BOLIVIA
BRAZIL
Paraná Basin
Campos Bsn
CHILE
Salta Group System
PARAGUAY
28°
Río Belén
Central System
Saliniana Bsn
ARGENTINA
Misiones Province
Santos Bsn
Pagan-cillo
Sierra Chica Córdoba
Ischigua-lasto
Western System
Marayes
Chaco-Paraná Bsn
Pelotas Bsn
Las Salinas
URUGUAY
Beazley
Villa Mercedes
Levalle Bsn
Atlantic Ocean
AFRICA
38°
Macachín-Quehué Bsn
Salado Bsn
120 Ma
Colorado Bsn
500 km

Areas with rifting and/or subsidence
Areas with active magmatism
Oceanic floor

◄ **Fig. 1.1** **a** Location of Sierra Chica of Córdoba and alkaline Cretaceous/Palaeogene provinces or localities peripheral to the tholeiitic lava flows of the Paraná basin, modified from Lagorio (2008). Cratons are shown with crosses on yellow background. Red area: volcanic outcrops of Paraná Magmatic Province; pink area: subsurface tholeiitic basalts of this igneous province in Argentina. Age of various volcanic events is shown in parentheses. **b** Areas of crustal extension and igneous activity during the Early Cretaceous in southernmost South America, taken from Uliana et al. (1990), Rossello and Mozetic (1999) and Ramos (1999). Bsn: basin

The proximity of the Etendeka locality to the Walvis Ridge, whose western end corresponds to the Tristan da Cunha seamount (composed of oceanic islands basalts, OIB), together with the closeness of the Paraná volcanism to the Rio Grande Rise (Fig. 1.2a), led some authors to conceive PEA LIP as the result of the upwelling of the so-called Tristan plume (e.g. Fodor 1987; O'Connor and Duncan 1990; Gibson et al. 2006). Other researchers explained the genesis of PEA LIP as a consequence of melting of the subcontinental lithospheric mantle, attributing to Tristan plume just a heat input (e.g. Piccirillo and Melfi 1988; Hawkesworth et al. 1992; Peate 1997; Marques et al. 1999). However, the Tristan plume has not been presently observed in seismic tomography (e.g. Montelli et al. 2006). Besides, Iacumin et al. (2003) considered that this volcanism was triggered by edge-driven convection caused by a significant change in the lithosphere thickness (King and Anderson 1995; King and Anderson 1998).

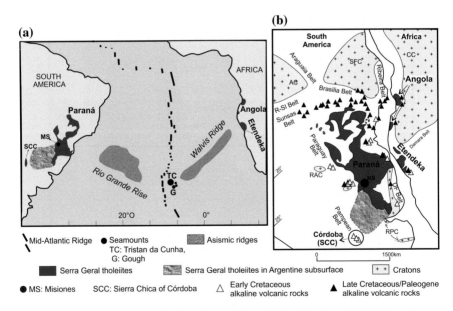

Fig. 1.2 **a** Schematic location of the lava flows of Paraná–Etendeka–Angola LIP. The mid-ocean ridge and seamounts are also represented. TC: Tristan da Cunha and G: Gough, simplified from O'Connor and Duncan (1990). **b** Illustration of the Paraná–Etendeka–Angola LIP including alkaline peripheral locations in the western sector of Gondwana, at approximately 130 Ma. Adapted from Piccirillo and Melfi (1988), Gibson et al. (1996), Marzoli et al. (1999), Lagorio (2008) and Lagorio and Vizán (2011). RPC: Río de la Plata craton, SFC: San Francisco craton, AC: Amazonian craton, CC: Congo craton, RAC: Rio Apa craton, R-SI: Rondonia-San Ignacio belt

Conversely, on the basis of recent Os isotopic data, Rocha-Júnior et al. (2012, 2013) proposed that only the sublithospheric mantle was involved in the generation of the PMP. These authors claimed that volcanism of PMP was determined by the combined effects of edge-driven convection and a large-scale mantle melting caused by the aggregation of continents to form the supercontinent of Pangea (Coltice et al. 2007) without requiring the involvement of a plume of any kind.

On the other hand, lower melting degrees of an OIB-like mantle and smaller lava volumes in SCC compared with those of the Paraná–Etendeka LIP have been previously considered by Kay and Ramos (1996) as a consequence of the large distance from the Tristan plume (Fig. 1.2a, b). Melting of subcontinental lithospheric mantle was afterwards invoked to explain SCC volcanism (Lucassen et al. 2002; Lagorio 2008).

In this contribution, it is analysed if both volcanic events, SCC and PMP, could have had different origins in the context of reorganization of geological plates during the Early Cretaceous. Considering a palaeogeographic reconstruction of Pangea for an age of 130 Ma, based on a new selection of palaeomagnetic poles of volcanic rocks from South America and Africa, different alternative models for the genesis of PMP and SCC volcanism are suggested.

References

Bellieni G, Brotzu P, Comin-Chiaramonti P, Ernesto M, Melfi AJ, Pacca G, Piccirillo EM (1984) Flood basalts to rhyolite suites in the Southern Paraná plateau (Brazil): paleomagnetism, petrogenesis and geodynamic implications. J Petrol 25(3):579–618

Cejudo Ruiz R, Goguitchaivili A, Geuna SE, Alva-Valdivia L, Solé J, Morales J (2006) Early cretaceous absolute geomagnetic paleointensities from Córdoba province (Argentina). Earth Planets Space 58(10):1333–1339

Chebli GA, Spalletti LA, Rivarola D, de Elorriaga E, Webster R (2005) Cuencas Cretácicas de la Región Central de la Argentina. Frontera Exploratoria de la Argentina. In: Chebli GA, Coriñas JS, Spalletti LA, Legarreta L, Vallejo EL (eds), 6º Congreso de Exploración y Desarrollo de Hidrocarburos. IAPG, Buenos Aires, pp 193–215

Coltice N, Phillips BR, Bertrand H, Ricard Y, Rey P (2007) Global warming of the mantle at the origin of flood basalts over supercontinents. Geology 35:391–394

Costa CH, Gardini CE, Chiesa JO, Ortiz Suárez AE, Ojeda GE, Rivarola DL, Tognelli GC, Strasser EN, Carugno Durán AO, Morla PN, Guerstein PG, Sales DA, Vinciguerra HM (2001) Hoja Geológica 3366-III, San Luis. Provincias de San Luis y Mendoza, vol 293. Instituto de Geología y Recursos Minerales, Servicio Geológico Minero Argentino. Boletín, Buenos Aires, p 67

Erlank AJ, Marsh JS, Duncan AR, Miller R McG, Hawkesworth CJ, Betton PJ, Rex DC (1984) Geochemistry and petrogenesis of the Etendeka volcanic rocks from SWA/Namibia. In: Erlank AJ (ed) Petrogenesis of the volcanic rocks of the Karroo Province, vol 13. Special Publication, Geological Society of South Africa, South Africa, pp 195–246

Ernesto M, Raposo MIB, Marques LS, Renne PR, Diogo LA, de Min A (1999) Paleomagnetism, geochemistry and $^{40}Ar/^{39}Ar$ dating of the North-Eastern Paraná Magmatic Province: tectonic implications. J Geodyn 28:321–340

Fodor RV (1987) Low- and high-TiO_2 flood basalts of southern Brazil: origin from a picritic parentage and a common mantle source. Earth Planet Sci Lett 84:423–430

Gibson SA, Thompson RN, Dickin AP, Leonardos OH (1996) Erratum to "High-Ti and low-Ti mafic potassic magmas: Key to plume-litosphere interactions and continental flood-basalt genesis". Earth Planet Sci Lett 141:325–341

Gibson SA, Thompson RN, Day JA (2006) Timescales and mechanism of plume lithosphere interactions: $^{40}Ar/^{39}Ar$ geochronology and geochemistry of alkaline igneous rocks from the Paraná-Etendeka large igneous province. Earth Planet Sci Lett 251:1–17

Gordillo CE (1972) Petrografía y composición química de los basaltos de la sierra de Las Quijadas —San Luis—y sus relaciones con los basaltos cretácicos de Córdoba. Boletín de la Asociación Geológica de Córdoba 1(3–4):127–129 (Córdoba)

Gordillo CE, Lencinas A (1967a) Geología y petrología del extremo norte de la Sierra de Los Cóndores, Córdoba. Boletín Academia Nacional de Ciencias 46(1):73–108 (Córdoba)

Gordillo CE, Lencinas A (1967b) El basalto nefelínico de El Pungo, Córdoba. Boletín Academia Nacional de Ciencias 46(1):109–115 (Córdoba)

Gordillo CE, Lencinas A (1969) Perfil geológico de la sierra Chica de Córdoba en la zona del río Los Molinos, con especial referencia a los diques traquibasálticos que la atraviesan. Boletín Academia Nacional de Ciencias 47:27–50 (Córdoba)

Hawkesworth CJ, Gallager K, Kelly S, Mantovani MSM, Peate D, Regelous M, Rogers N (1992) Parana magmatism and the opening of the South Atlantic. In: Storey B, Alabaster A, Pankhurst R (eds) Magmatism and causes of Continental break- up, vol 68. Geological Society Special Publication, London, pp 221–240

Iacumin M, De Min A, Piccirillo EM, Bellieni G (2003) Source mantle heterogeneity and its role in the genesis of Late Archean-Proterozoic (2.7–1.0 Ga) and Mesozoic (200 and 130 Ma) tholeiitic magmatism in the South American Platform. Earth Sci Rev 62:365–397

Janasi VA, Freitas VA, Heaman LH (2011) The onset of flood basalt volcanism, Northern Parana basin, Brazil: U-Pb baddeleyite/zircon age for a Chapeco-type dacyte. Earth Planet Sci Lett 302:147–153

Kay SM, Ramos VA (1996) El magmatismo cretácico de las sierras de Córdoba y sus implicancias tectónicas. 13° Congreso Geológico Argentino y 3° Congreso de Exploración de Hidrocarburos, vol 3. Actas, Buenos Aires, pp 453–464

King SD, Anderson DL (1995) An alternative mechanism of flood basalt formation. Earth Planet Sci Lett 136:269–279

King SD, Anderson DL (1998) Edge-driven convection. Earth Planet Sci Lett 160:289–296

Kuiper KF, Deino A, Hilgen FJ, Krijgsman W, Renne PR, Wijbrans JB (2008) Synchronizing rock clocks of Earth history. Science 320:500–504

Lagorio SL (2008) Early Cretaceous alkaline volcanism of the Sierra Chica de Córdoba (Argentina): Mineralogy, geochemistry and petrogenesis. J S Am Earth Sci 26:152–171

Lagorio SL, Vizán H (2011) El volcanismo de Serra Geral en la Provincia de Misiones: aspectos geoquímicos e interpretación de su génesis en el contexto de la Gran Provincia Ígnea Paraná-Etendeka-Angola. Su relación con el volcanismo alcalino de Córdoba (Argentina). Geoacta 36:27–53

Linares E, Valencio DA (1974) Edades Potasio-Argón y paleomagnetismo de los diques traquibasálticos del río de Los Molinos, Córdoba, República Argentina. Revista de la Asociación Geológica Argentina 29(3):341–348

Lucassen F, Escayola MP, Romer RL, Viramonte JG, Koch K, Franz G (2002) Isotopic composition of Late Mesozoic basic and ultrabasic rocks from the Andes (23–32°S)—implications for the Andean mantle. Contrib Miner Petrol 143:336–349

Marques LS, Dupre B, Piccirillo EM (1999) Mantle source compositions of the Parana Magmatic Province (Southern Brazil): evidence from trace element and Sr-Nd–Pb isotope geochemistry. J Geodyn 28:439–458

Martínez AN, Rivarola D, Strasser E, Giambiagi L, Rouquet MB, Tobares ML, Merlo M (2012) Petrografía y geoquímica preliminar de los basaltos cretácicos de la sierra de Las Quijadas y cerrillada de Las Cabras, provincia de San Luis, Argentina. Serie Correlación Geológica 28(1): 9-22. Aportes al Magmatismo y Metalogenia Asociada I, Tucumán

Marzoli A, Melluso L, Morra V, Renne PR, Sgrosso I, D'Antonio M, Duarte Morais L, Morais EAA, Ricci G (1999) Geochronology and petrology of Cretaceous basaltic magmatism in the Kwanza basin (Western Angola), and relationships with the Paraná-Etendeka continental flood basalt province. J Geodyn 28:341–356

Montelli R, Nolet G, Dahlen FA, Masters G (2006) A catalogue of deep mantle plumes: New results from finite-frecuency tomography. Geochem Geophys Geosyst 7(11):1–69

O'Connor JM, Duncan RA (1990) Evolution of the Walvis Ridge-Rio Grande rise hot spot system: Implications for African and South American plate motions over plumes. J Geophys Res 95 (B11):17475–17502

Peate DW (1997) The Paraná-Etendeka Province. In: Mahoney JJ, Coffin MF (eds) Large igneous provinces: continental oceanic and planetary flood volcanism, vol 100. Geophysical Monograph American Geophysical Union, Boulder, Colorado, pp 215–245

Piccirillo EM, Melfi AJ (1988) The mesozoic flood volcanism from the Paraná basin (Brazil): petrogenetic and geophysical aspects. Universidad de São Paulo, San Pablo, p 600

Ramos VA (1999) Evolución tectónica de la Argentina. Instituto de Geología y Recursos Minerales, Geología Argentina, Anales 29(24):715–784, Buenos Aires

Renne PR, Ernesto M, Pacca IG, Coe RS, Glen JM, Prévot M, Perrin M (1992) The age of Paraná flood volcanism, rifting of Gondwanaland, and Jurassic-Cretaceous boundary. Science 258:975–979

Renne PR, Deckart K, Ernesto M, Feraud G, Piccirillo EM (1996) Age of the Ponta Grossa dike swarm (Brazil) and implications to Paraná flood volcanism. Earth Planet Sci Lett 144:199–211

Rocha-Júnior ERV, Puchtel IS, Marques LS, Walker RJ, Machado FB, Nardy AJR, Babinski M, Figueiredo AMG (2012) Re–Os isotope and highly siderophile element systematics of the Parana Continental Flood Basalts (Brazil). Earth Planet Sci Lett 337–338:164–173

Rocha-Júnior ERV, Marques LS, Babinski M, Nardy AJR, Figueiredo AMG, Machado FB (2013) Sr-Nd–Pb isotopic constraints on the nature of the mantle sources involved in the genesis of the high-Ti tholeiites from northern Parana Continental Flood Basalts (Brazil). J S Am Earth Sci 46:9–25

Rossello E, Mozetic ME (1999) Caracterización estructural y significado geotectónico de los depocentros cretácicos continentales del centro—oeste argentino. Boletim do 5º Simpósio sobre o Cretáceo do Brasil, UNESP—Campus de Rio Claro/SP: 107–113

Thiede DS, Vasconcelos PM (2010) Paraná flood basalts: Rapid extrusion hypothesis confirmed by new $^{40}Ar/^{39}Ar$ results. Geology 38(8):747–750

Uliana MA, Biddle KT, Cerdan J (1990) Mesozoic extension and the formation of Argentine sedimentary basins. In: Tankard AJ, Balkwill HR (eds) Extensional tectonics and stratigraphy of the North Atlantic margins. American Asociation of Petroleum Geologists, Memoir 46:599–614, Tulsa

Webster RE, Chebli GA, Fischer FJ (2004) General Levalle basin, Argentina: A frontier Lower Cretaceous rift basin. American Asociation of Petroleum Geology Bulletin 88(5):627–652. Tulsa, Oklahoma

Wildner W, Santos JOS, Hartmann LA, McNaughton NJ (2006) Clímax final do vulcanismo Serra Geral em 135 Ma: primeiras idades U-Pb em zircão. 43º Congresso Brasileiro Geologia, Extended Abstracts, Aracaju

Chapter 2
Early Cretaceous Volcanism in Central Argentina

Abstract The main exposed site of Early Cretaceous volcanism in central Argentina is located in Sierra Chica of Córdoba Province (SCC), within the Central Rift System. Also to the south, in Levalle basin, a thick Early Cretaceous volcanic pile lies buried in the subsurface. Other localities where volcanism is exposed are Sierra de las Quijadas and Cerrillada de las Cabras of San Luis Province. In SCC, as in the other mentioned localities, a volcanic–sedimentary complex was developed under rifting tectonics. Lava flows are frequently associated with scoria fall, pyroclastic and phreatomagmatic breccias within a strombolian-type volcanism. A new $^{40}Ar/^{39}Ar$ dating performed on sanidine phenocrysts of a trachyte from Almafuerte locality indicated an age of 129 ± 1 Ma. Diverse groups of rocks, mainly of potassic character, were distinguished: (1) alkali basalt—trachyte suite, (2) transitional basalt—latibasalt suite, (3) basanite—phonolite suite and (4) ankaratrites. Magma evolution must have taken place at crustal level(s) from distinct parental melts, mainly through fractional crystallization in an open-system magma chamber. Mantle source composition supports residual garnet and phlogopite, it does not exhibit features related to slab-derived metasomatism despite its location over Pampean mobile belt, and bears a lithospheric nature. SCC volcanism is of high Ti, display similarities with potassic Brazilian localities around Paraná basin as Alto Paranaíba and Goias, pointing out analogies in their mantle sources.

Keywords Early Cretaceous · Sierra Chica · Córdoba · Alkali basalts · Transitional basalts · Basanites · Ankaratrites · Lithospheric mantle

2.1 Early Cretaceous Volcanism in Córdoba Province

2.1.1 Volcanism of the Sierra Chica of Córdoba Province

2.1.1.1 Geological Setting

Córdoba ranges are essentially integrated by three mountain belts called Sierra Norte, Sierra Grande and Sierra Chica (Fig. 2.1), in the geological context of the

© The Author(s) 2016
S.L. Lagorio et al., *Early Cretaceous Volcanism in Central and Eastern Argentina During Gondwana Break-Up*, SpringerBriefs in Earth System Sciences, DOI 10.1007/978-3-319-29593-0_2

9

eastern Sierras Pampeanas. Early Cretaceous alkaline volcanism is located in Sierra Chica. By contrast, Late Cretaceous, very small volcanic outcrops appear in Sierra Grande (Estancia Guasta and Ciénaga Grande) and to the south, in the locality of Chaján (Fig. 2.1).

Córdoba ranges are mainly composed of igneous–metamorphic basement of Precambrian–Early Carboniferous age, formed essentially by biotite gneisses, schists, cordierite migmatites, granitic bodies, pegmatites and aplites, with minor participation of amphibolites, ultramafic rocks and marbles (e.g. Gordillo and Lencinas 1980; Martino et al. 1995; Kraemer et al. 1995; Geuna et al. 2008; Lira and Sfragulla 2014).

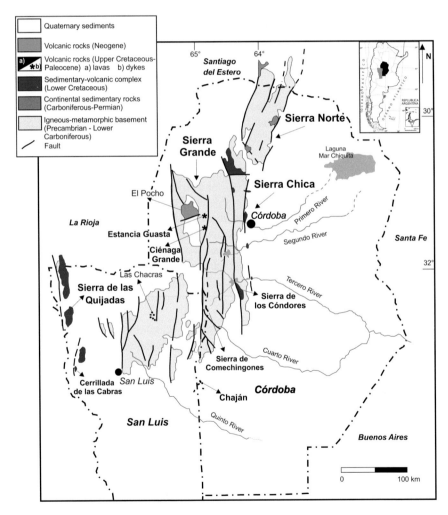

Fig. 2.1 Geological map of the provinces of Córdoba and San Luis, adapted from Lucero Michaut et al. (1995) and Caminos and González (1996)

Over the basement, the Early Cretaceous volcanic–sedimentary complex lies, generated as a consequence of extensional conditions (Uliana et al. 1990), within the Central Rift System (Rossello and Mozetic 1999; Ramos 1999; Fig. 2.2a). The structural style was mainly controlled by the Punilla and La Calera faults, with a dominant NNW direction (Fig. 2.2b). These structures were developed in zones of crustal weakness (Schmidt et al. 1995; Martino et al. 2014), as the suture represented by the eastern ophiolite belt of Kraemer et al. (1995) between the para-autochthonous Córdoba terrane and the Río de la Plata craton. Therefore, the rift could have been developed in the hanging wall of the suture between amalgamated terranes (Kay and Ramos 1996, Ramos et al. 2000). Andean tectonics in the area that belongs to the Nazca flat slab, provoked the lifting of Sierra Chica, exposing the Cretaceous complex through reverse faults that were previously normal faults (Sisto et al. 1993; Schmidt et al. 1995; Kay and Ramos 1996; Martino et al. 2014).

Outcrops of the Lower Cretaceous volcanic rocks in Sierra Chica are represented by remnants of a volcanic–sedimentary complex unconformably overlying its basement. This complex is composed of continental red beds and alkaline basic lavas (Gordillo and Lencinas 1980) that belong to syn-rift deposits of basins located from the south of Sierra Chica to Sierras de Guasayán (in Santiago del Estero Province, to the north of Córdoba Province) in a transtensional context (Schmidt et al. 1995). The kinematics of the associated strike-slip faults has been mentioned in different papers. Several authors indicate sinistral lateral displacements (Sisto and Cortés 1992; Sisto et al. 1993, 1995; Sánchez et al. 1995; Kay and Ramos 1996), whereas others consider dextral movements (Schmidt et al. 1995; Martino et al. 2014). These latter authors have pointed out that a dextral rifting would have determined the formation of depocenters as isolated pull-apart basins during the Early Cretaceous. This is consistent with the interpretation of the strike-slip faults in Paraná and Colorado basins pointed out by Uliana et al. (1990) and Tankard et al. (1995). The filling of the Lower Cretaceous basins of Sierra Chica would have taken place in restricted half-grabens. The sediments belong to alluvial fans, braided rivers and ephemeral lake deposits (e.g. Poiré et al. 1989; Sánchez et al. 1990; Astini et al. 1993; Piovano 1996). The climate must have been of a semi-arid and oxidizing environment, because there are no fossil records available. Four sedimentary depocenters were identified, three of them with considerable thicknesses. They remained as remnants after intense Tertiary erosion. The largest deponcenters are located in the following: (1) Sierra de Masa, Sierra de Copacabana and Sierra del Pajarillo, (2) the Saldán locality and (3) Sierra de los Cóndores (southern portion of Sierra Chica), whereas the smaller one is located in a zone known as El Pungo (4) (Fig. 2.2b, c). An updated and detailed sedimentological and stratigraphical study of the deposits that filled these depocenters has been recently presented by Astini and Oviedo (2014).

The sedimentary beds are interbedded with lava flows in the mountains of Sierra de los Cóndores and El Pungo; in the Saldán locality, only conglomerates with clasts of basalts and hydrothermal minerals have been recognized (Piovano 1996). Schmidt et al. (1995) considered two megasequences of Cretaceous deposits in Sierra Chica (Fig. 2.2c). The lower one reaches the largest thickness in Sierra de Copacabana and

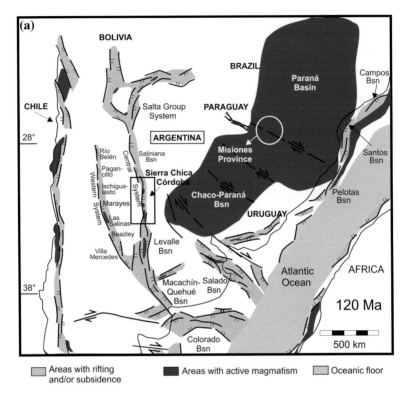

Fig. 2.2 a Extensional tectonics during Early Cretaceous (taken from Uliana et al. 1990), showing Sierra Chica of Córdoba as part of the Central Rift System (Rossello and Mozetic 1999; Ramos 1999). **b** Geological map of the Sierra Chica, based on Gordillo and Lencinas (1980), Schmidt et al. (1995) and Kay and Ramos (1996), as presented by Lagorio (2008). **c** Synthetic scheme of the main Cretaceous successions in the area of Sierra Chica, taken from Schmidt et al. (1995); division in megasequences proposed by these authors based on the presence of volcanism, with later modifications introduced by Minudri and Sánchez (1994) and Piovano (1996)

Sierra del Pajarillo, and there are no lava flows interbedded with the sedimentary deposits. The volcanic levels outcrop in El Pungo, Sierra de los Cóndores and around the localities of Almafuerte and Despeñaderos localities (Fig. 2.2b); in the area of Los Molinos dam, only several basaltic dykes that could belong to the Early Cretaceous volcanic event are recorded intruding the crystalline basement.

2.1.1.2   Volcanism in different localities of Sierra Chica

Sierra de los Cóndores

This range belongs to the southern end of Sierra Chica (Fig. 2.2b). It is composed of a volcanic–sedimentary complex that reaches over than 250 m unconformably overlying crystalline basement rocks.

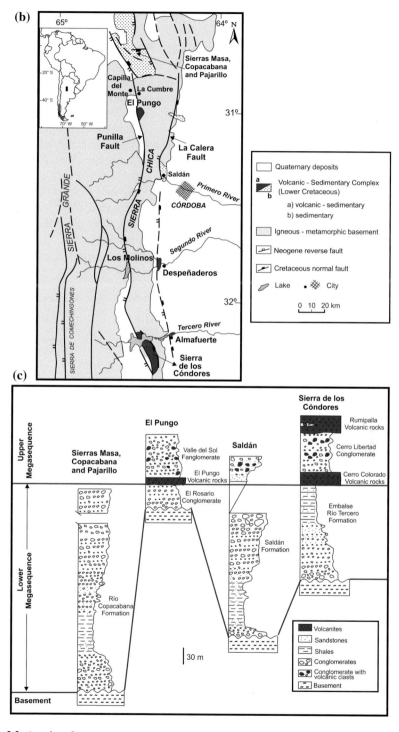

Fig. 2.2 (continued)

The volcanic rocks of this range were previously described by Tannhauser (1906), Bodenbender (1907) and Pastore (1930). Gordillo and Lencinas (1967a) defined in the northern section of this range the Sierra de los Cóndores Group, composed of a basal sedimentary unit (Embalse Río Tercero Formation) which is followed by two volcanic cycles (Cerro Colorado and Rumipalla Volcanic rocks, respectively) separated by a sedimentary unit (Cerro Libertad Conglomerate).

Gordillo and Lencinas (1967a) considered that a non-foidic trachybasalt rich in sanidine was the dominant rock type of the area; other varieties (picritic basalts, calc-alkaline and alkaline trachytes, rhombic porphyries) were subordinate and restricted to local lava flows. The rocks form a series of moderate alkaline character, relatively enriched in potassium.

Although radiometric dating coincides throughout the entire range (Table 2.1), Poiré et al. (1989), Sánchez et al. (1990) and Ferreira Pittau et al. (2008) considered that the volcanism must have begun in the southern area, whereas in the northern sector, there was an intense sedimentation. In the southern sector, the volcanic activity was characterized by a significant pyroclastic component, within a context

Table 2.1 Previous radiometric ages for volcanic rocks from Sierra Chica de Córdoba (SCC)

Locality	Rock type/mineral phase	K/Ar Age (Ma)	Corrected K/Ar Age (Ma)	Ref.
Sierra de los Cóndores				
Cerro Colorado	Trachybasalt	120	120	1
Cerro Colorado	Trachybasalt	117	117	1
Cerro Colorado	Trachybasalt	120 ± 5	123 ± 5	2
Cerro Colorado (mean age)	Trachybasalt	116 ± 5	119 ± 5	2
Cerro Colorado (mean age)	Trachybasalt	118.5 ± 6	121.5 ± 6	3
Cerro Libertad	Trachyte	128 ± 5	128 ± 5	1
Cerro Quebracho	Trachybasalt	116 ± 6	119 ± 6	3
Gianna quarry (Berrotarán)	Sanidine	112 ± 6	115 ± 6	1
Berrotarán	Sanidine	120 ± 2	123 ± 2	4
Berrotarán	Sanidine	130 ± 6	133 ± 6	4
El Quebracho quarry	Alkaline basalt	117 ± 5	120 ± 5	6
Almafuerte				
Middle section	Trachybasalt	122 ± 3	125 ± 3	4
Middle section	Trachybasalt	129 ± 8	132 ± 8	4
Upper section	Andesite	119 ± 10	122 ± 10	5
Upper section	Basalt	121 ± 5	124 ± 5	5
Dyke	Trachyte	120 ± 2	123 ± 2	4
Close to Río Tercero	Basalt	123 ± 4	126 ± 4	7
Two km to the N of Almafuerte	Trachybasalt	124 ± 5	127 ± 5	1
Close to Almafuerte locality	Porphyry	114 ± 5	117 ± 5	5

(continued)

Table 2.1 (continued)

Locality	Rock type/mineral phase	K/Ar Age (Ma)	Corrected K/Ar Age (Ma)	Ref.
Los Molinos dam				
Dyke N° 8 (mean age)	Trachybasalt	150 ± 10	154 ± 10	8
Dyke N° 5 (mean age)	Trachybasalt	140 ± 10	143 ± 10	8
Dyke N° 10 (mean age)	Trachybasalt	138 ± 10	141 ± 10	8
Dyke N° 9 (mean age)	Trachybasalt	131 ± 10	134 ± 10	8
Dyke N° 12 (mean age)	Trachybasalt	129 ± 10	132 ± 10	8
Dyke N° 4 (mean age)	Trachybasalt	68 ± 5	70 ± 5	8
Dyke N° 13 (mean age)	Trachybasalt	65 ± 5	67 ± 5	8
Dyke N° 14 (mean age)	Trachybasalt	63 ± 5	65 ± 5	8
Dyke N° 14	Trachybasalt	60.2 ± 3	61.7 ± 3	9
El Pungo				
El Pungo-La Cumbre	Basalt	119 ± 5	122 ± 5	4
El Pungo-La Cumbre	Basalt	119	119	10
Despeñaderos				
Saldán Formation	Basalt	122.9 ± 1.3	122.9 ± 1.3	11
Saldán Formation	Basalt	136.3 ± 1.5	136.3 ± 1.5	11
Saldán Formation	Basalt	130.7 ± 1.4	130.7 ± 1.4	11
Saldán Formation	Basalt	130.4 ± 1.4	130.4 ± 1.4	11
Saldán Formation	Basalt	128.5 ± 1.5	128.5 ± 1.5	11

Data source (1) Gordillo and Lencinas (1967a); (2) Stipanicic and Linares (1969); (3) Valencio (1972); (4) González and Kawashita (1972); (5) Linares and González (1990); (6) Cortelezzi et al. (1981); (7) Valencio and Vilas (1972); (8) Linares and Valencio (1974); (9) Gordillo and Lencinas (1969); (10) Lencinas (1971); (11) Cejudo Ruiz et al. (2006). A correction coefficient of 1.025 was applied to datings perfomed prior to 1978 (INGEIS 1977)

of strombolian-type volcanism, with scoriaceous and cinder cones together with the eruption of lava flows (Ferreira et al. 1999).

Towards the southern sector, the thickness of the volcanic rocks notably increases (Bodenbender 1929; Pensa 1957; Ferreira et al. 1999), and there are frequent volcanic dykes as feeders of lava flows with thickness that reach about 100 m (Sánchez and Bermúdez 1997). Escayola et al. (1998) described the roots of a volcanic neck that must have been an important centre of eruption lava and pyroclastic material in the southern part of this mountain chain. Sánchez and Bermúdez (1997) presented a map of the whole range, also distinguishing the units Cerro Colorado and Rumipalla Volcanic rocks. They noted three volcanic cycles, the last two integrating the Rumipalla Volcanic rocks.

Geuna (1997, 1998), Lagorio et al. (1997) and Lagorio (2003) performed sampling at different sections in order to complement previous studies of Gordillo and Lencinas in Sierra de Los Cóndores, carrying out an integrated palaeomagnetic and petrological–geochemical study. Figures 2.3 and 2.4 show some of the typically plateau-shaped hills, where several samplings have been performed. In these

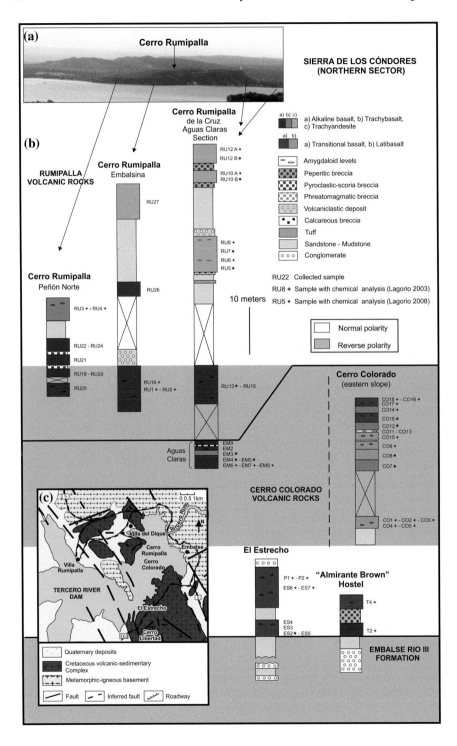

◀ **Fig. 2.3** Sierra de los Cóndores (northern sector). **a** Panoramic view of Cerro Rumipalla from El Estrecho, as shown in Lagorio (2003) and Lagorio et al. (2014). **b** Stratigraphic sections around the Tercero River dam, with a stratigraphic proposal according to the recorded magnetic polarities (Geuna 1998), including the lithological types; volcanic rocks classified according to their chemistry. The subdivision in units defined by Gordillo and Lencinas (1967) for Sierra de los Cóndores Group is also indicated. **c** Schematic geological map based on Sánchez et al. (1990)

profiles, the diverse lithological types were also identified (Figs. 2.3b and 2.4b, d), according to the geochemical classification and the petrographic features, as will be described in the following item.

The sampled levels of volcanic rocks are massive, black to dark brown, with auburn lava flows that bear amygdales of granular zeolites and carbonate reflecting a deuterism sometimes intense.

Sedimentary and volcanic strata are very frequently interbedded in the northern sector, which indicates the simultaneity between both processes.

In some outcrops, the overlapping of lava flows produced marked thermal effects. In other cases, however, peperitic breccias were generated; these are characterized by fragments often with crenulated edges, invaded by the fine sandy matrix. Although the term breccia is used here, subrounded fragments predominate, which characterizes the globular, fluidal or pillowy peperites (Busby-Spera and White 1987). The genesis of deposits with peperitic texture can be due to lava flowing on wet unconsolidated sediments or due to the falling of pyroclasts in a plastic state (e.g. glassy shards) on a material with those characteristics. This is consistent with the presence of cord and pillow lavas on the southern shore of the lake formed by the Tercero River dam, mentioned by Sánchez and Bermúdez (1997), who interpreted this as an evidence of a possible subaqueous eruption. Furthermore, levels corresponding to gravitational flows, debris flow type, produced by remobilization of volcanic material are also observed. Outcrops corresponding to scoria fall and pyroclastic breccia deposits are particularly abundant in the southern sector of the range (Fig. 2.4d), which constitute typical products of strombolian activity, characterized by moderate explosiveness.

Volcanic breccias are constituted by fragments of vesicular lava generally with sharp edges, contained in a lapilli or tuffaceous matrix; sometimes, the fragments are cemented by carbonate or analcime (Fig. 2.4f). Volcanic bombs are also observed in these deposits, particularly in the southern area as pointed out by Ferreira et al. (1999). It should be noted that also some phreatomagmatic episodes must have also taken place locally. For example, a lapilli-bearing level mostly composed of angular fragments from the basement along with subordinate juvenile fragments has been observed in the northern sector (Cerro Colorado, Fig. 2.3b; "hybrid volcanic rock" of Gordillo and Lencinas 1967a).

On the other hand, also in the northern zone (the Aguas Claras area) a level of a calcrete, partially bearing a texture of a breccia, has been recognized, with similarities to the lithology described by Piovano (1994) in the locality of Saldán zone. This latter author attributes its genesis to thermal groundwater, inherent to the volcanic process, in an arid context with high evaporation and low biological activity.

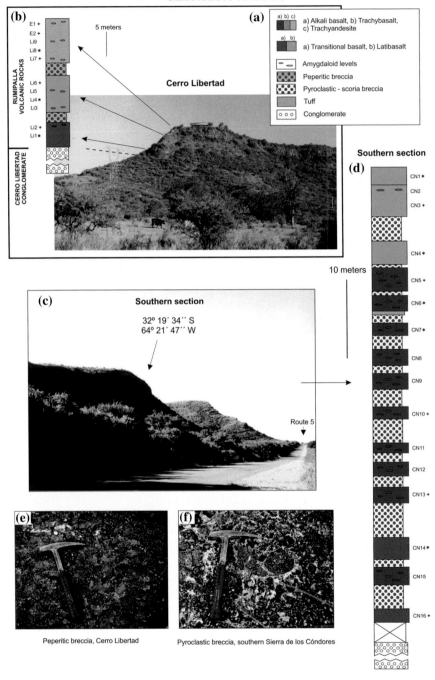

SIERRA DE LOS CÓNDORES

(a)

a) b) c) a) Alkali basalt, b) Trachybasalt, c) Trachyandesite

a) b) a) Transitional basalt, b) Latibasalt

Amygdaloid levels

Peperitic breccia

Pyroclastic - scoria breccia

Tuff

o o o Conglomerate

(b)

RUMIPALLA VOLCANIC ROCKS

E1 +
E2 +
Li9
Li8 *
Li7 +
Li6 +
Li5
Li4 *
Li3
Li2 +
Li1 *

CERRO LIBERTAD CONGLOMERATE

5 meters

Cerro Libertad

Southern section

(d)

CN1 *
CN2
CN3 +
CN4 *
CN5 +
CN6 *
CN7 *
CN8
CN9
CN10 +
CN11
CN12
CN13 +
CN14 *
CN15
CN16 +

10 meters

(c)   **Southern section**

32° 19´ 34´´ S
64° 21´ 47´´ W

Route 5

(e)

(f)

Peperitic breccia, Cerro Libertad

Pyroclastic breccia, southern Sierra de los Cóndores

◄ **Fig. 2.4** Sierra de los Cóndores (southern section, south of the dam), as shown in Lagorio (2003) and Lagorio et al. (2014). **a** View of Cerro Libertad from El Estrecho. **b** Stratigraphic section in this hill, including different lithological types according to their chemistry; the subdivision of Sierra de los Cóndores Group defined by Gordillo and Lencinas (1967) is also indicated. **c** View of the outcrops in the southern sector of Sierra de los Cóndores. **d** Stratigraphic section including the different lithological types depending on their chemistry. Note that from Cerro Libertad to the south, all lavas were extruded in a normal magnetic polarity, different from what is recorded in the northern sector (area of the Tercero River dam). Pensa (1957) suggested that Cerro Libertad was the boundary between the northern and southern areas of Sierra de los Cóndores. **e** Detail of peperitic breccia in Cerro Libertad. **f** Detail of pyroclastic breccia in southern Sierra de los Cóndores

Petrological–geochemical and palaeomagnetic studies allowed establishing a stratigraphic proposal for the northern sector of the range (Fig. 2.3b). For this area, the lava flows are distributed throughout the whole sequence, diachronically and with lateral variations between different localities. According to the recorded magnetic polarities (Geuna 1998), the time span covering the volcanic event does not exceed 3 million years (Geuna 1997; Lagorio 2003). In this context, the distinction between two different volcanic cycles which can be mapped seems to be extremely difficult.

If the lava flows of the Libertad hill and those of the southern section are compared, it appears that both sites complete their evolution in one chron of normal polarity (Fig. 2.4) which is different to what happens in the northern sector (Geuna 1998). From the Libertad hill to the south, all the lava flows were extruded during a normal polarity chron. Previously, Pensa (1957) placed in this hill the boundary between the southern and northern areas of this range. Anyway, it is not possible to determine in which of the two normal polarity intervals recorded in the northern sector were extruded the lavas in the southern area, and if they poured out synchronously.

The locality of Almafuerte

The volcanic–sedimentary complex is exposed to about 3 km from the town of this name, near the Piedras Moras dam on the Tercero River, approximately 15 km to the east of Sierra de los Cóndores (Figs. 2.2b and 2.5). This succession is composed of lava flows interbedded with sandstones and breccias that are cut by dykes; it has has a thickness of 160 m, tilts 25°–30° to the SSE (Schröder 1967; Mendía 1978) and was entirely observed prior to the building of the dam. Presently, only the upper part of the section is outcropping.

Gordillo and Lencinas (1980) considered that these outcrops belong to the Sierra de los Cóndores Group, whereas Piovano (1996) includes this sequence, like others located to the east of the La Calera fault, in the Saldán Formation.

Geuna (1997) and Lagorio (2003) described the sequence close to Route 36, north of the bridge over the Tercero River (Fig. 2.5a, b). This is a sequence formed by several lava flows, with maximum thickness of 4 m, interbedded with breccia and sandstone levels of minor thickness, cut by basic dykes (Fig. 2.5c–e). Breccia levels are of different types.

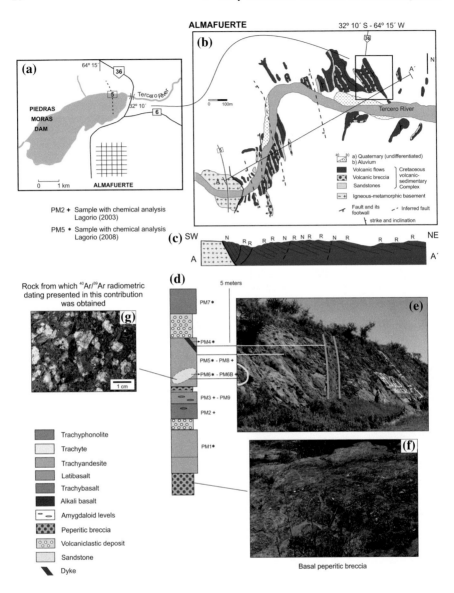

ALMAFUERTE   32° 10´ S - 64° 15´ W

PM2 + Sample with chemical analysis
       Lagorio (2003)

PM5 * Sample with chemical analysis
       Lagorio (2008)

Rock from which ⁴⁰Ar/³⁹Ar radiometric
dating presented in this contribution
was obtained

Trachyphonolite

Trachyte

Trachyandesite

Latibasalt

Trachybasalt

Alkali basalt

Amygdaloid levels

Peperitic breccia

Volcaniclastic deposit

Sandstone

Dyke

Basal peperitic breccia

A peperitic-type breccia is located in the lower part of the section, characterized by large fragments of volcanic rocks with marked crenulations and interpenetrations of a matrix composed of reddish fine sandstone (Fig. 2.5f). It is noteworthy that Pensa (1957) pointed out a very frequent interaction between lava flows and sediments in this area, assigning special importance to the type of wetting affecting the sands, as a significant factor to determine the final texture of rocks. In fact, the degree of saturation in water, the consolidation and permeability of the sediments must have strongly conditioned the lithological features of the deposits. Other

◄ **Fig. 2.5** Locality of Almafuerte, as shown in Lagorio (2003) and Lagorio et al. (2014). **a** Location map of this area with the former track of Route 5 (now submerged). **b** Geological map of the area for times prior to the dam construction (based on Schröder 1967 and Mendía 1978); the area of the stratigraphic section is indicated (see track of Route 5 for reference). **c** Schematic section showing the magnetic polarity, as determined by Mendía (1978) and reinterpreted by Geuna (1997); note that at most three alternate polarities are present at each structural block. **d** Stratigraphic section of the currently outcropping levels close to the Piedras Moras dam, including the lithological types according to their chemistry. The magnetic polarity of this section is reverse, while basaltic dyke PM4 is of normal polarity. **e** Photograph of the lava flows (trachyandesite) intruded by a basaltic dyke (sample PM4); this lava flow presents also sectors of trachytic composition (sample PM6). **f** Photograph of basal peperitic breccias. **g** Photograph of the trachyte (PM6) composed of macrophenocrysts and abundant phenocrysts of feldspar; the study under the microscope allowed the identification of anorthoclase with a thick rim of sanidine constituting the macrophenocrysts, whereas phenocrysts are essentially of sanidine. The $^{40}Ar/^{39}Ar$ of 129.6 ± 1 Ma dating was obtained in these phenocrysts

breccia levels unconformable overlain lava flows, indicating a brief erosive pre-depositional stage; they show subangular fragments without any interpenetration by thin sandstone matrix. They are interpreted as the result of debris flows. Schröder (1967) assigned this genesis to all of the breccia levels of this area.

A macroporphyritic trachyte-bearing sanidine crystals (Fig. 2.5g) was selected for dating through $^{40}Ar/^{39}Ar$ techniques, as it will be explained below. According to palaeomagnetic data from the entire sequence, including levels studied by Mendía (1978) before the water reservoir was filled, Geuna (1997) pointed out that most of the lava flows would have extruded during three chrones of alternate polarity (R–N–R, Fig. 2.5c). Meanwhile, the dykes must have intruded the section during a second period of normal polarity. Then, volcanic eruptions in the Almafuerte locality and northern Sierra de los Cóndores (area of the Tercero River dam) show a similar pace and are probably coeval.

The locality of Despeñaderos

The Cretaceous volcanic–sedimentary complex outcrops in parts of the margins of the Segundo River, near the Despeñaderos locality (Fig. 2.2b). It constitutes isolated outcrops in some canyons of the river, as they were observed by Bain Larrahona (1940), Kull and Methol (1979), and Gordillo and Lencinas (1980) who included them in the Sierra de Los Cóndores Group; on the other hand, Piovano (1996) included these outcrops in the Saldán Formation.

Geuna (1997) and Lagorio (2003) described the outcrops located west of the intersection with Route 36, about 300 m from the bridge over the Segundo River, on its right margin. Outcrops consist of lava flows interbedded with breccia levels making up a sequence of about 25 m, subhorizontal, originated during only one chron of normal polarity (Fig. 2.6a, b).

Previously, Gordillo and Lencinas (1969) analysed the petrography and chemistry of a sample from the left margin of the Segundo River, here also included in the classification diagrams.

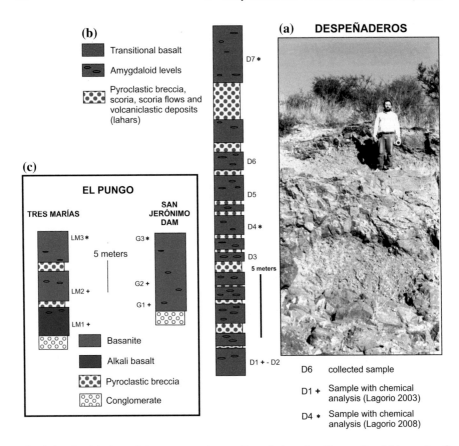

Fig. 2.6 a Partial view of the outcrops close to Despeñaderos locality, on the right margin of Segundo River, and to the west of the intersection with Route 36. Note the alternation of lava flows and breccia levels that form the sequence. **b** Stratigraphic section pointing out the lithological types according to their chemistry. **c** Stratigraphic section in El Pungo locality, note the lithological types according to their chemistry. Modified from Lagorio et al. (2014)

The sampled basaltic levels have thicknesses of up to 5 m, reddish-brown colour and are markedly amygdaloidal, denoting an intense deuterism under highly oxidizing conditions. The breccia levels have thicknesses between 0.20 and 3 m (Fig. 2.6b). Pyroclastic breccias, scoria fall deposits as well as dense gravitational flows (debris flows) of laharic type are also recorded, suggesting an origin related to a volcanic area with high and, therefore, unstable slopes.

An arid to semi-arid climate context with torrential rains and low vegetation must have favoured the formation of debris flows, as described by Piovano and Astini (1990) for basal conglomeratic facies of the Saldán Formation that integrate thick alluvial fan deposits, approximately at 60 km to the north of the Despeñaderos locality.

The locality of Los Molinos

In this area, basic dykes that intrude the crystalline basement are outcropping (Figs. 2.2b). Gordillo and Lencinas (1969) described 35 dykes, with thicknesses between 0.5 and 1.5 m, length of the order of a hundred metres, subvertical and generally NNW trending.

They were discriminated by Gordillo and Lencinas (1969) as unaltered dykes (olivine trachydiabases) and hydrothermalized (analcimites and potassium trachydiabases). These authors stated that despite having obtained a K–Ar age of 60.2 ± 3 Ma, the intrusive phases of Los Molinos must belong to the Early Cretaceous magmatic event, considering the mineralogical and chemical similarity to the lavas of Sierra de los Cóndores.

While later radiometric dating obtained by Linares and Valencio (1974) suggested two groups of contrasting ages (150 ± 10 Ma–129 ± 10 Ma and 68 ± 5 Ma–63 ± 5 Ma), Geuna (1997) and Geuna and Vizán (1998) determined that their palaeomagnetic directions are all reversed and similar to those of the lavas of Sierra de los Cóndores, which reinforces the hypothesis of a single intrusive event taking place during the Early Cretaceous.

Lagorio (2003) collected samples from several dykes in the proximity of the dam to characterize them from a petrological and geochemical point of view. While Lagorio (2008) presented complete chemical data for samples of dykes 4 and 14 (olivine trachydiabases of Gordillo and Lencinas 1969), Lagorio (2003) also reported chemical analyses of samples of dykes 8 and 9 (olivine analcimites of Gordillo and Lencinas 1969), characterized by high loss on ignition values conferred by the presence of analcime. For this contribution, the chemistry of the two varieties was considered. Also, an analysis of the potassic trachydiabase of Gordillo and Lencinas (1969) was even included in the classification diagrams presented below.

El Pungo—La Cumbre localities

In the northern sector of Sierra Chica, close to the towns of La Cumbre and Capilla del Monte (Fig. 2.2b), the outcrops of the Cretaceous volcanic–sedimentary complex have been defined as El Pungo Group by Gordillo and Lencinas (1967b). These authors described this group as formed by a thick basal conglomerate (El Rosario Conglomerate) on which successively lay a nephelinic basalt (El Pungo Volcanic) and sedimentary deposits (Valle del Sol Fanglomerate).

Later, Delpino et al. (1999), Sánchez et al. (1999, 2001) and Sánchez (2001a, b) defined the formational units El Rosario, El Saucecito and Peñón Blanco, as members of the El Pungo Group, excluding the Valle del Sol Fanglomerate. El Saucecito Formation is the volcanic unit that includes the lava flows described by Gordillo and Lencinas (1967b) and also the volcaniclastic deposits. The latter are more abundant than the volcanic rocks and related to hydroclastic and strombolian eruptions (Delpino et al. 1999; Sánchez et al. 2001, 2002). The volcanic flows were studied also by Ancheta et al. (2002).

Geuna (1997) and Lagorio (2003) performed studies in three sites: (1) San Jerónimo dam, (2) Tres Marías town and (3) Estancia El Rosario area. The volcanic sequence has a thickness of less than 20 m. The lava flows are covering conglomerates (Fig. 2.6c); usually they are massive, only in some sectors they have carbonate amygdales, and seem to have been erupted through two chrones of normal and reverse polarity. In the area of Tres Marías, the lava flows are between 3 and 5 m thick and are interbedded by thin breccia/lapillitic levels that usually do not exceed 2 m in thickness (Fig. 2.6c). These levels have subangular and angular fragments of basalt, of various sizes, that belong to both lapilli and blocks. They were interpreted as fall deposits in a volcanic strombolian context, relatively close to the centre of eruption. Close to this area, Delpino et al. (1999) studied thick pyroclastic rocks that Sánchez et al. (2001) and Ancheta et al. (2002) characterized as hydromagmatic types and considered that they were formed in the initial periods of the volcanic activity in this region.

More recently, Oviedo and Astini (2014) presented a detailed study along Road El Cuadrado, across the Sierra Chica, that allowed the identification and characterization of a volcaniclastic unit that seems equivalent to that described by Delpino et al. (1999) and Sánchez et al. (1999) in El Pungo region. Oviedo and Astini (2014) better considered to preserve the original nomenclature of Gordillo and Lencinas (1967b), though amending it, in order to include in El Pungo Formation other materials. These comprise primary fall deposits, possible surge ones, gravity flows as well as reworked deposits, along with basaltic flows. Those authors consider that those deposits represent relicts of a syneruptive strombolian and phreatomagmatic activity close to a basaltic monogenic system conforming the synrift filling of the Cretaceous depocenters.

2.1.1.3   Age of this Volcanism

This volcanism has been intensively dated, mainly several decades ago, through the K/Ar method with ages ranging from 154 ± 10 Ma to 61.7 ± 3 Ma (Table 2.1). The latest K/Ar whole-rock dating on rocks from the Despeñaderos locality ranges from 136.3 to 122.9 Ma (Cejudo Ruiz et al. 2006).

The vast majority of dating was performed on whole-rock samples, so that frequent partially crystallized glassy groundmass, subtle deuteric alteration and weathering, as well as a possible insufficient outgassing during the volcanic event, could have determined the large age variability. Nevertheless, a slight diachronism considering the whole Sierra Chica de Córdoba cannot be ruled out. It should be noted, however, that no more than three periods of alternate magnetic polarity were recorded in every section according to the Early Cretaceous reversal rate that would mean a maximum time of 3 million years (My) in continuous records. Also, the presence of magnetic reversals reinforces the fact that the volcanic event in SCC occurred mostly before 124 Ma (the onset of the Cretaceous Normal SuperChron).

We made a detailed microscopic analysis of several rock types from different localities of SCC, previously to the selection of samples. $^{40}Ar/^{39}Ar$ dating was then performed on a sample collected at the Almafuerte locality (32° 10'S; 64° 14'W), as it bears numerous sanidine phenocrysts. This sample corresponds to a trachyte with 8.62 wt% K_2O which was measured at Actlabs (Canada), through step-heating analysis. It should be noted that biotite (though scarce) only appears in this sample of SCC, while phlogopite was not even reported in any rock. The sample wrapped in Al foil was loaded in evacuated and sealed quartz vial with K and Ca salts and packets of LP-6 biotite used as a flux monitor. It was irradiated in the nuclear reactor for 48 h. After the flux monitors were run, J values were then calculated for the sample, using the measured flux gradient. LP-6 biotite has an assumed age of 128.1 Ma. The neutron gradient did not exceed 0.5 % on sample size. The Ar isotope composition was measured in a Micromass 5400 static mass spectrometer. 1200 °C blank of ^{40}Ar did not exceed $n \times 10^{-10}$ cc STP.

The sample yielded an age spectrum with six steps plateau characterized by 89 % of ^{39}Ar. Therefore, an age value of 129.6 ± 1.0 Ma was obtained, as it is shown in Fig. 2.7a.

On the Inverse Isochrone Plot Plateau points form a linear trend characterized by age value of 128.4 ± 1.5 Ma, with a mean square weighted deviation (MSWD) = 1.6 and $(^{40}Ar/^{36}Ar)_0$ = 348 ± 27 (Fig. 2.7b). Inverse isochrone age (IIA) is concordant with weighted mean plateau age (WMPA) in the frame of error. Thus, WMPA, as more precise, should correspond to the age of formation or closing of isotope system of the sanidine.

2.1.1.4   Classification and Petrography

The first detailed studies on the petrography, mineralogy and geochemistry of Cretaceous volcanic rocks in Sierra Chica were made by Gordillo and Lencinas (1967a, b, 1969). Subsequently several authors conducted studies in different areas of Sierra Chica: Cortelezzi et al. (1981), Kay and Ramos (1996), Bermúdez and Sánchez (1997), Lagorio et al. (1997), Lagorio (1998), Escayola et al. (1998, 1999), Lucassen et al. (2002) and Ancheta et al. (2002). Lagorio (2003, 2008) carried out a petrological and geochemical study on these volcanic rocks throughout the Sierra Chica, including mineralogical analyses using electron microprobe.

The results are summarized below, considering also samples analysed by other authors. The classification was performed based on the chemistry of the major elements. The samples analysed in this study present a continuous variation from basalt to trachyte in the diagram alkalis versus silica (TAS, Le Bas et al. 1986; Fig. 2.8a) including trachybasalts, basaltic trachyandesites and trachyandesites. A few number of samples were classified in the fields of basanite and basaltic andesite.

The diagram of classification R1-R2 (De la Roche et al. 1980; Bellieni et al. 1981; Fig. 2.8b) allows the recognition of two volcanic suites: (1) alkali basalt–

Fig. 2.7 a $^{40}Ar/^{39}Ar$ age and Ca/K spectra for sanidine phenocrysts of a trachyte of Sierra Chica of Córdoba (SCC). The sample yielded age spectrum with six steps plateau characterized by 89 % of ^{39}Ar, age value of 129 ± 1.0 Ma (WMPA). **b** On the Inverse Isochrone Plot Plateau points form a linear trend characterized by the age value of 128.4 ± 1.5 Ma, MSWD = 1.6 and $(^{40}Ar/^{36}Ar)_0 = 348 ± 27$. *WMPA* weighted mean plateau age; *MSWD* mean square weighted deviation; *TFA* total fusion age; *IIA* inverse isocrone age

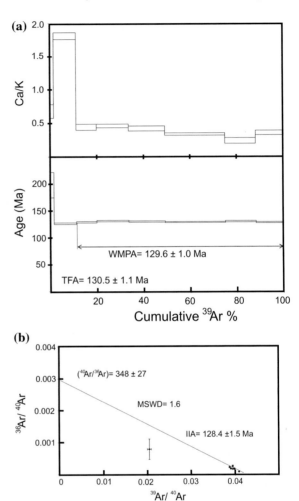

trachybasalt–trachyandesite–trachyphonolite–trachyte (AkB-Tc), and (2) transitional basalt–latibasalt (TrB-Lb), as well as two groups of rocks: basanites (Bsn) and ankaratrites (Akr). The latter present samples with elevated loss on ignition (>5 %), determined by abundant analcime that characterizes these rocks (analcimites of Gordillo and Lencinas 1969). However, classification based on immobile trace element ratios (e.g. Floyd and Winchester 1978) of the few samples with high loss on ignition included here confirm classification through major elements. It should be noted that, considering that the diagram R1-R2 allows a more complete discrimination of these rocks, it was thus adopted for the nomenclature.

The basanites and part of the rocks of the alkaline suite are CIPW-normative nepheline (Ne); meanwhile, the transitional suite is characterized, instead, for being hypersthene normative (Hy). Furthermore, the nomenclature employed by Gordillo and Lencinas (1967a) for some samples of Sierra de los Cóndores (e.g. calc-alkaline

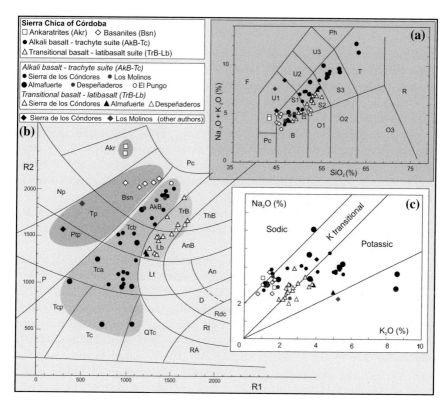

Fig. 2.8 Classification diagrams of volcanic rocks of Sierra Chica, adapted from Lagorio (2008) and Lagorio et al. (2014). The following data are also included: potassic trachydiabase of Los Molinos and alkali basalt of Despeñaderos according to analysis presented by Gordillo and Lencinas (1969); basanite and phonotephrite from the southern sector of Sierra de Los Cóndores provided by Cortelezzi et al. (1981) and Escayola et al. (1998). **a** Alkalis versus silica diagram (TAS, Le Bas et al. 1986); *B* basalt, *S1* trachybasalt, *S2* basaltic trachyandesite, *S3* trachyandesite, *T* trachyte, *F* foidite, *U1* basanite/tephrite, *U2* phonotephrite, *U3* tephriphonolite, *Ph* phonolite, *O1* basaltic andesite. **b** R1-R2 diagram (from De la Roche et al. 1980; Bellieni et al. 1981); R1 = 4Si − 11 (Na + K) − 2 (Fe + Ti), R2 = 6Ca + 2Mg + Al; *Pc* picrite, *Akr* ankaratrite, *Bsn* basanite, *AkB* alkali basalt, *TrB* transitional basalt, *ThB* tholeiitic basalt, *Np* nephelinite, *Tp* tephrite, *Tcb* trachybasalt, *Lb* latibasalt, *AnB* andesi-basalt, *Ptp* phonotephrite, *Tca* trachyandesite, *Lt* latite, *P* phonolite, *Tph* trachyphonolite, *T* trachyte, *QTc* quartz trachyte, *Lan* latiandesite, *An* andesite, *QLc* quartz latite, *D* dacite, *Rd* rhyodacite, *R* rhyolite, *AR* alkaline rhyolite. **c** Na$_2$O versus K$_2$O diagram with fields according to Comin-Chiaramonti et al. (1997)

trachytes, equivalents of latibasalts in the classification used here) let infer a transitional character that reinforces the distinction between two suites.

According to the criteria of alkalinity of Comin-Chiaramonti et al. (1997), the analysed rocks have essentially a potassic (1 < K$_2$O/Na$_2$O ≤ 2) and transitional to potassic (Na$_2$O − 2 < K$_2$O and K$_2$O/Na$_2$O ≤ 1) character; only few examples typify as high potassic rocks (K$_2$O/Na$_2$O > 2) or as sodic lithological types (Na$_2$O − 2 ≥ K$_2$O), as it is shown in Fig. 2.8c.

Alkaline Suite (BA-T)

Alkali basalts outcrop in Sierra de los Cóndores (there, they are comparable to the picritic basalts and partially to the brown andesinic trachybasalts of Gordillo and Lencinas 1967a), Almafuerte and Los Molinos (olivine trachydiabases of Gordillo and Lencinas 1969). They present a porphyritic texture, with phenocrysts and microphenocrysts of olivine (Fo_{90-83}) and clinopyroxene ($Wo_{53-46}En_{46-35}Fs_{15-5}$); the glassy groundmass consists of clinopyroxene ($Wo_{51-49}En_{43-41}Fs_{8-6}$), olivine ($Fo_{87-79}$), Ti-magnetite (ulvöspinel = 63–48 %) ± ilmenite (R_2O_3 = 7 %), alkali feldspar ($Or_{62-12} Ab_{69-37}An_{26-1}$) and plagioclase ($An_{30-26}$). Apatite and Ti-biotite occur as accessory minerals. The most primitive rocks usually have small ultramafic xenoliths (lherzolitic, dunitic, harzburgitic, Fig. 2.9a).

On the other hand, some samples of Los Molinos dykes exhibit quartz xenocrysts with rims of reaction formed by clinopyroxene; they also have granitic basement xenoliths.

There are frequent ocellar-type textures, particularly in Sierra de los Cóndores, defining ellipsoidal patches that consist of sanidine and/or nepheline, egirinic clinopyroxene ± Ti-biotite ± apatite (Fig. 2.9b).

In relation to dykes from Los Molinos locality, according to their chemical composition, they classify as alkaline basalts and based on petrographic features they were considered as lamprophyres by Gordillo and Lencinas (1969). Particularly, they typify as hyalomonchiquites (IUGS nomenclature: Le Maitre 1989; Le Bas and Streckeisen 1991). However, the absence of biotite and/or amphibole phenocrysts does not allow, in fact, considering them as classical lamprophyres. These phases appear only in the groundmass in small proportions and could have even been formed by devitrification.

The sample collected by Gordillo and Lencinas (1969) near the Despeñaderos locality, also included in the classification diagrams, typifies as alkali basalt (Fig. 2.8b), and the petrographic features described by those authors are similar to those described here for this lithotype.

Trachybasalts outcrop in Sierra de los Cóndores (equivalent to black andesinic trachybasalts of Gordillo and Lencinas 1967a) and the Almafuerte locality. Two varieties, one porphyritic and other aphyric, are recognized. The first consists of phenocrysts and/or microphenocrysts of clinopyroxene ($Wo_{48}En_{46-45}Fs_{8-6}$), olivine ($Fo_{86-72}$) ± plagioclase ± Ti-magnetite immersed in a hypocrystalline basis formed by alkali feldspar ($Or_{58-39}Ab_{47-34}An_{14-8}$), plagioclase ($An_{46}$), clinopyroxene ($Wo_{46}En_{47}Fs_7$), olivine ($Fo_{72}$), Ti-magnetite (Ulv. = 59 %) ± ilmenite (R_2O_3 = 7 %), apatite and Ti-biotite. Occasionally, few xenocrysts of plagioclase and quartz with reaction rims and schist xenoliths are observed. Ocellar-type textures are frequent, defining ellipsoidal patches composed of sanidine and/or nepheline, aegirinic clinopyroxene ± Ti-biotite ± apatite. The aphyric trachybasalts (Sierra de los Cóndores, Fig. 2.9c) consist of clinopyroxene crystals ($Wo_{49-46}En_{46-41}Fs_{10-6}$), olivine ($Fo_{77}$), alkali feldspar ($Or_{67-42}Ab_{49-30}An_{9-3}$) and plagioclase ($An_{45}$); feldspars are characterized by including small crystals of

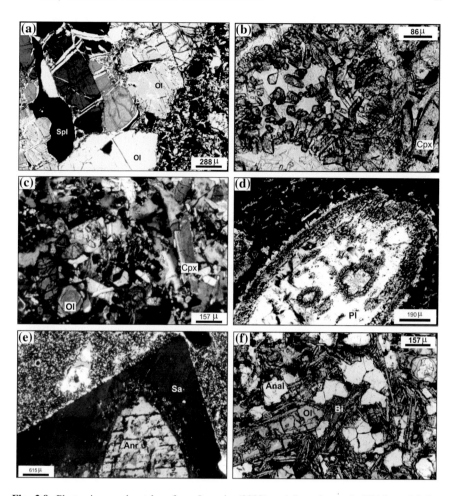

Fig. 2.9 Photomicrographs, taken from Lagorio (2003) and Lagorio et al. (2014). **a** Olivine crystals (Ol_{89}) showing straight contacts and interstitial spinel section (Spl) (picotite variety) forming lherzolitic xenolith in alkali basalt (picritic variety of Gordillo and Lencinas 1967a) from Aguas Claras (northern Sierra de los Cóndores), crossed nicols, 3.5X. **b** Ocellar structure in alkali basalt from the north of Sierra de los Cóndores) formed by egirinic clinopyroxene and alkali feldspar; note the compositional contrast with the microphenocrysts located in the lower right corner (*chestnut colour*), parallel nicols, 20X. **c** Trachybasalt from the north of Sierra de los Cóndores; crystals of olivine (Ol), clinopyroxene (Cpx), and opaque minerals are observed encompassed by feldspar sections, crossed nicols, 10X. **d** Sieve texture in macrophenocrysts of plagioclase (Pl) of a trachyandesite from the northern sector of Sierra de los Cóndores (rhombic porphyry of Gordillo and Lencinas 1967a), crossed nicols, 10X. **e** Macrophenocryst of anorthoclase (Anr) with large euhedral rim of sanidine (Sa) immersed in fine bostonitic groundmass, belonging to the trachyte of Almafuerte locality, crossed nicols, 2.5X. **f** Crystals of olivine (Ol) replaced by bowlingite, analcime (Anl), biotite (Bt), clinopyroxene and of opaque minerals along with interstitial glass forming an analcimite (ankaratrite), corresponding to one of the dykes of Los Molinos, parallel nicols, 10X

clinopyroxene, altered olivine and Ti-magnetite (Ulv. = 50 %) $\pm$ ilmenite (R_2O_3 = 6 %).

Trachyandesites outcrop in Sierra de los Cóndores (alkali trachytes of Gordillo and Lencinas 1967a) and the Almafuerte locality. Whereas in the first locality, they are Hy-normative, in the second one, Ne- and quartz-normative varieties occur. They show from microporphyritic to macroporphyritic textures characterized by phenocrysts and/or microphenocrysts of clinopyroxene ($Wo_{50-44}En_{48-40}Fs_{14-6}$), iddingsitized olivine $\pm$ Ti-magnetite and feldspar (as long as 1 cm). In macroporphyritic rocks, there are plagioclase crystals (An_{51-46}) as well as anorthoclase ($Or_{28-23}Ab_{48-45}An_{29-27}$) ones, in both Hy-normative (e.g. rhombic porphyry of Gordillo and Lencinas 1967a from Sierra de los Cóndores) and Q-normative (the rocks of the Almafuerte locality) varieties. In the latter, anorthoclase crystals are covered by a coat of sanidine, and also biotite and kaersutite phenocrysts as well as apatite microphenocrysts occur. In all of these varieties, crystals of feldspar and clinopyroxene show disequilibrium textures (Fig. 2.9d) with frequent resorption, strong corrosion and sieve texture. Groundmass is composed of alkali feldspar ($Or_{59-25}Ab_{52-39}An_{27-2}$),   plagioclase   ($An_{52-31}$),   clinopyroxene ($Wo_{48-43}En_{48-42}Fs_{10-8}$), iddingsitized olivine, titanomagnetite (Ulv. = 84–4 %) $\pm$ ilmenite (R_2O_3 = 8–7 %), with apatite and biotite as accessory phases.

Trachytes outcrops in Almafuerte are Q-normative and have macroporphyritic texture (Fig. 2.5g). They are composed of crystals of 15 mm of anorthoclase ($Or_{22}Ab_{63}An_{15}$) covered by a thick edge of sanidine ($Or_{47}Ab_{47}An_6$) (Fig. 2.9e), phenocrysts and microphenocrysts of sanidine ($Or_{53-49}Ab_{43}An_4$), plagioclase (An_{54-53}), biotite and clinopyroxene ($Wo_{50-48}En_{45-43}Fs_{8-6}$), with microphenocrysts of Ti-magnetite and apatite. Groundmass is composed of sodic sanidine ($Or_{62}Ab_{36}An_2$), clinopyroxene ($Wo_{48}En_{44}Fs_8$), biotite, Ti-magnetite (Ulv. = 52 %), apatite and $\pm$ quartz in sparse fine veinlets and microamygdales.

Trachyphonolites also outcrop in the locality of Almafuerte. They are Ne-normative and have macroporphyritic texture. They are characterized by having plagioclase crystals (An_{52-45}) up to 10 mm long, microphenocrysts and phenocrysts of clinopyroxene ($Wo_{48-47}En_{43-42}Fs_{11-9}$) along with microphenocrysts of olivine altered to iddingsite, Ti-magnetites and apatite, immersed in a fine-grained matrix. The latter is made up of plagioclase, clinopyroxene, Ti-magnetite (Ulv. = 78 %), iddingsitized olivine, biotite and apatite. Plagioclase and clinopyroxene crystals show strong corrosion and sieve textures.

Some rocks of the suite bear vesicles filled by an intermediate term of the series analcime-wairakite (Lagorio and Montenegro 1996) along with calcite and smectites.

It should be noted that alkali basalts, trachybasalts, trachyphonolites and some trachyandesites are Ne-normative (Ne up to 7.30 % in trachyphonolites). Most trachyandesites are Hy-normative (Hy between 2.90 and 12.31 %) and a few Q-normative (Q $\sim$ 1 %); on the other hand, the latter feature is dominant in trachytes (Q upto 7.88 %).

Transitional Suite

Transitional basalts outcrop in Sierra de los Cóndores (equivalent to the brown and black andesinic trachybasalts of Gordillo and Lencinas 1967a) and in the locality of Despeñaderos. They have both porphyritic and aphyric textures. The first consists of phenocrysts and microphenocrysts of olivine altered to iddingsite together with clinopyroxene microphenocrysts ($Wo_{50-48}En_{46-41}Fs_{10-7}$). The glassy groundmass contains clinopyroxene ($Wo_{49}En_{45}Fs_6$), iddingsitized olivine, plagioclase (An_{29-26}), alkaline feldspar and Ti-magnetite (Ulv. = 81 %); accessory phases such as apatite and biotite are common. The aphyric variety consists of olivine crystals altered to iddingsite, plagioclase (An_{43}), alkali feldspar ($Or_{36-16}Ab_{69-55}An_{14-9}$) and clinopyroxene; feldspars enclose smaller crystals of clinopyroxene ($Wo_{45}En_{47}Fs_8$), iddingsitized olivine, Ti-magnetite and apatite.

Latibasalts are rocks from Sierra de los Cóndores (calc-alkaline trachytes of Gordillo and Lencinas 1967a) and the locality of Almafuerte. They have porphyritic texture defined by phenocrysts of iddingsitized olivine and clinopyroxene ($Wo_{49-41}En_{51-42}Fs_{10-5}$), together with occasional plagioclase phenocrysts and microphenocrysts of Ti-magnetite. The crystals of clinopyroxene and plagioclase often show sieve texture. The mesostasis consists of clinopyroxene ($Wo_{50-42}En_{50-43}Fs_{8-7}$), olivine altered to iddingsite, Ti-magnetite (Ulv. = 70–59 %), plagioclase (An_{48-28}), alkali feldspar ($Or_{47-33}Ab_{63-52}An_{4-1}$), apatite and occasional biotite. Some xenocrysts of quartz with reactions rims formed by small crystals of clinopyroxene are sporadically observed.

Rocks of the transitional suite are Hy-normative (Hy up to 15.75 %).

Basanites

Basanites outcrop in El Pungo locality; they have porphyritic texture composed of phenocrysts and microphenocrysts of olivine (Fo_{87}) and clinopyroxene ($Wo_{52}En_{44-42}Fs_{6-5}$). The glassy groundmass is made up of clinopyroxene ($Wo_{51}En_{44}Fs_5$), olivine (Fo_{87}), Ti-magnetite (Ulv. = 86 %) and sodic anorthoclase ($Or_{11}Ab_{76}An_{13}$), together with apatite and Ti-biotite as accessory phases. Dunitic xenoliths and glomeroporphyritic clusters of olivine sparsely occur. They possess high content of Ne-normative (Ne between 8.20 and 14.48 %).

Studies performed by Cortelezzi et al. (1981) and Escayola et al. (1998) in the southern section of Sierra de los Cóndores revealed the presence of a basanitic–tephritic to tephriphonolitic lavas in Basalto and Quebracho quarries. In the first site, Cortelezzi et al. (1981) recognized a monzodioritic diabase, whereas in the second locality, a nephelinitic basanite. Subsequently, rocks of the latter type were described by Escayola et al. (1998) also in Basalto quarry, as part of the exposed roots of a volcanic neck bearing numerous mantle xenoliths that represents an important centre of effusion of lavas in the southern part of the range. According to the description presented by those authors, the rocks have porphyritic texture with phenocrysts of olivine and clinopyroxene, immersed in a groundmass made up of nepheline and augite microlites, Ti-magnetite and scarce biotite, involving also sanidine which is not associated with nepheline and shows an irregular arrangement. Rocks bear numerous xenocrysts of olivines as well of xenoliths of garnet and spinel lherzolites, dunites and alkali feldspar syenites. There are also abundant pegmatoid segregations of alkali gabbroid composition. Xenoliths of garnet lherzolite are the first ones reported from all the alkaline Cretaceous lithological types of the central–south-western area of South America (Escayola et al. 1998). According to chemical data presented by the aforementioned authors, samples of the Basalto and Quebracho quarries plot in the fields U1 (basanite/tephrite) and U2 (phonotephrite) in TAS diagram, whereas in the diagram R1-R2, they typify as basanite and phonotephrite, respectively (Fig. 2.8a, b). Lucassen et al. (2002) presented several samples classifying in those fields also from that zone. More recently, Ferreira Pittau et al. (2008) defined an eruptive cycle of basanite–phonolite composition in the southern sector of Sierra de los Cóndores.

The sample from Los Molinos locality characterized as a potassic trachydiabase by Gordillo and Lencinas (1969) classify as tephrite in both TAS and R1-R2 diagrams (Fig. 2.8a, b). It was described by those authors as a porphyritic rock with olivine phenocrysts altered to antigorite and/or calcite, phenocrysts of diopsidic augite and Ti-biotite, set in a groundmass rich in sanidine including microlites of plagioclase and Ti-magnetite, bearing calcite nodules and veins of analcime.

Ankaratrites

These rocks belong to the dykes that outcrop in the locality of Los Molinos locality that were considered by Gordillo and Lencinas (1969) as olivinic analcimites. Microscopic studies show that they are composed of clinopyroxene, analcime, olivine altered to saponite, Ti-biotite, kaersutite, opaque minerals, carbonate, apatite, interstitial glass replaced by clays of the group of smectites, little alkali feldspar and small sections of euhedral sphene, which form a compact texture (Fig. 2.9f). These rocks were considered as lamprophyres by Gordillo and Lencinas (1969) and typified as analcimic monchiquites on the basis of the classification proposed by the IUGS. However, they have not the porphyritic texture characteristic of this type of rocks.

2.1.1.5 Mineralogy

Lagorio (2003, 2008) presented microprobe analyses that were carried out by a Cameca–Camebax operating at 15 kV and 15 nA at the Dipartimento di Mineralogia e Petrologia of Padova University (Italy). The PAP-Cameca program has been used to convert X-ray counts into weight percentage of the corresponding oxides. Results are considered accurate within 2–3 % for major elements and 9 % for minor elements. Selected analyses of mineral phases are given in different tables, expanded from those presented in Lagorio (2008), while the complete set of data is reported in Lagorio (2003).

Clinopyroxene

Selected clinopyroxene compositions of SCC volcanic rocks are given in Table 2.2. They plot as diopside and augite (Fig. 2.10a) in the Ca–Mg–Fe* (Fe^{2+} + Fe^{3+} + Mn) diagram of Morimoto et al. (1988).

Clinopyroxenes show high values of Mg, so that a clear trend towards Fe* is not defined, as in other alkaline provinces (Fig. 2.10a).

Clinopyroxenes of basanites show the highest Ca content (up to 50.64 %), whereas those from the transitional suite reach the lowest Ca values (Table 2.2; Fig. 2.10b). It is also noteworthy the slight core to rim increase in Ca in some phenocryst and/or microphenocrysts, and especially the slight but generalized Mg increase from crystal rims towards groundmass microlites (Table 2.2; Fig. 2.10c).

TiO_2 is variable (0.14–4.71 wt%), usually lower than 3 % (mean 1.63 ± 0.74 %) so that SCC clinopyroxenes are not titaniferous according to Deer et al. (1992), except some late-crystallized ones from Los Molinos (MO1) and Almafuerte (PM4) AkB dykes. Ti/Al ratios of all the SCC clinopyroxenes range from 0.2 to 0.5, as expected for moderately alkaline rocks, as it should be taken into account that Ti/Al ratios > 0.5 are typical of strongly alkaline to peralkaline rocks (Cundari and Ferguson 1982).

Early-crystallized clinopyroxenes display Al^{VI} ranging from 0.018 to 0.091 a.f.u. (mean 0.038 ± 0.024 a.f.u.), pointing out, in general, low-pressure crystallization (Nimis 1995; Nimis and Ulmer 1998). The highest crystallization pressures have been found (Nimis and Ulmer 1998) for the Cpx phenocryst of the alkali basalt PM4 (Table 2.2), i.e. core = 4.7 and rim = 5.7 kbar, given the relatively high Al^{VI} content (0.091 and 0.098, respectively).

Olivine

Selected olivine compositions of SCC volcanic rocks are given in Table 2.3. The forsterite (Fo) content of olivine pheno- and microphenocrysts of basanites and alkali basalts ranges from 90–84 % (core) to 89–83 % (rim), and from 87 to 79 % for groundmass olivine (microlites). Similarly, the olivine of trachybasalts spans from Fo 86–75 % (core) to 85–71 % (rim) and 72 % (microlites).

Comparison between mg# [Mg/(Mg + Fe^{2+})] values of olivines and those [Mg/(Mg + Fe^{2+}); Fe_2O_3/FeO = 0.18] of host rocks shows that some early-crystallized olivines may have mg# values quite lower than those expected, suggesting crystallization from Mg poorer melts.

Table 2.2 Clinopyroxene microprobe compositions from selected rocks of Sierra Chica de Córdoba, expanded from Lagorio (2008)

Sample	G3			MO1			EM5		EM3		
Suite/group	Bsn			AkB-Tc			AkB-Tc		AkB-Tc		
Rock type	Bsn			AkB			AkB		Tcb		
Locality	El Pungo			Los Molinos			Cóndores		Cóndores		
	mp/c	mp/r	gm	mp/c	mp/r	gm	gm	gm	mp/c	mp/r	gm
SiO_2	50.50	49.80	50.79	51.95	44.93	48.92	51.79	48.78	51.42	51.23	51.33
TiO_2	2.06	2.29	1.81	1.35	4.71	2.66	1.88	2.61	1.15	1.48	0.95
Al_2O_3	4.07	5.01	3.31	2.04	6.74	4.07	2.76	4.88	2.34	2.48	1.98
FeO_{total}	5.49	5.24	4.76	5.15	9.63	6.43	5.78	6.14	5.36	6.28	4.77
MnO	0.13	0.00	0.16	0.19	0.13	0.09	0.13	0.11	0.09	0.17	0.07
MgO	14.82	14.02	15.02	15.9	10.98	14.27	15.14	13.78	15.83	15.51	15.98
CaO	24.33	24.19	24.66	22.76	22.15	22.85	23.84	23.73	22.72	22.97	21.90
Na_2O	0.44	0.44	0.30	0.37	0.69	0.41	0.41	0.48	0.39	0.35	0.34
Cr_2O_3	0.27	0.60	0.44	0.28	0.04	0.30	0.14	0.38	0.13	0.06	0.47
Sum	102.10	101.59	101.26	100.00	100.00	100.00	101.87	100.89	99.43	100.51	97.79
$Fe_2O_3^*$	3.10	1.84	1.89	1.14	2.84	1.81	1.47	3.07	2.22	1.77	0.83
Si	1.825	1.812	1.851	1.910	1.696	1.815	1.878	1.791	1.898	1.883	1.924
Al (IV)	0.173	0.188	0.142	0.089	0.300	0.178	0.118	0.209	0.102	0.107	0.076
Sum	1.998	2.000	1.993	1.999	1.996	1.993	1.996	2.000	2.000	1.990	2.000
Al (VI)	0.000	0.027	0.000	0.000	0.000	0.000	0.000	0.002	0.000	0.000	0.011
Fe^{2+}	0.082	0.109	0.093	0.127	0.223	0.149	0.135	0.104	0.104	0.144	0.126
Fe^{3+}	0.084	0.050	0.052	0.032	0.081	0.005	0.040	0.085	0.062	0.049	0.023
Cr	0.008	0.017	0.013	0.008	0.001	0.009	0.004	0.011	0.004	0.002	0.014
Mg	0.798	0.760	0.816	0.871	0.618	0.789	0.818	0.754	0.871	0.850	0.893
Mn	0.004	0.000	0.005	0.006	0.004	0.003	0.004	0.003	0.003	0.005	0.002
Ti	0.056	0.063	0.050	0.037	0.134	0.074	0.051	0.072	0.032	0.041	0.027
Ca	0.942	0.943	0.963	0.896	0.896	0.908	0.926	0.934	0.898	0.904	0.879
Na	0.031	0.031	0.022	0.027	0.005	0.003	0.029	0.034	0.028	0.025	0.025
Sum	2.005	2.000	2.014	2.004	2.007	2.012	2.007	1.999	2.002	2.020	2.000
Ca	49.32	50.64	49.92	46.38	49.18	47.81	48.15	49.68	46.34	46.31	45.71

(continued)

Table 2.2 (continued)

Sample	G3			MO1			EM5	EM3		
Suite/group	Bsn			AkB-Tc			AkB-Tc	AkB-Tc		
Rock type	Bsn			AkB			AkB	Tcb		
Locality	El Pungo			Los Molinos			Cóndores	Cóndores		
	mp/c	mp/r	gm	mp/c	mp/r	gm	gm	mp/c	mp/r	gm
Mg	41.78	40.82	42.30	45.08	33.92	41.55	40.11	44.94	43.55	46.44
Fe*	8.90	8.54	7.78	8.54	16.90	10.64	10.21	8.72	10.14	7.85

Sample	RU5					RU10B	EM5	EM3			
Suite/group	AkB-Tc					AkB-Tc	AkB-Tc	AkB-Tc			
Rock type	Tca					Tca	AkB	Tcb			
Locality	Cóndores					Cóndores	Cóndores	Cóndores			
	p/c	p/r	mp/c	mp/r	gm	p/c	p/r	mp1/c	mp1/r	mp2/c	mp2/r
SiO_2	51.41	51.73	49.60	49.17	52.41	50.32	49.99	51.56	51.37	50.34	50.29
TiO_2	0.89	0.71	2.21	2.55	1.04	1.54	1.78	1.34	1.32	1.58	1.56
Al_2O_3	3.39	2.79	4.71	4.57	2.81	4.14	5.32	3.56	3.4	4.19	4.2
FeO_{total}	9.37	8.67	8.60	8.39	6.40	9.81	9.32	5.93	6.08	9.84	10.04
MnO	0.44	0.40	0.20	0.17	0.17	0.26	0.21	0.05	0.12	0.27	0.17
MgO	13.15	13.26	14.57	13.26	16.44	13.93	13.99	16.08	16.19	13.8	13.36
CaO	21.85	21.94	21.09	22.97	20.59	20.4	20.07	21.56	21.57	20.14	20.61
Na_2O	1.14	0.83	0.79	0.50	0.58	0.76	0.7	0.54	0.55	0.65	0.71
Cr_2O_3	0.00	0.00	0.08	0.00	0.68	0.02	0.08	0.57	0.76	0.00	0.00
Sum	101.65	100.34	101.84	101.58	101.13	101.19	101.47	101.19	101.34	100.81	100.96
Fe_2O_3*	4.73	2.40	4.62	3.17	1.49	3.32	2.28	2.13	2.91	2.12	2.47
Si	1.878	1.915	1.802	1.805	1.902	1.848	1.828	1.869	1.86	1.859	1.857
Al (IV)	0.122	0.085	0.198	0.195	0.098	0.152	0.172	0.131	0.140	0.141	0.143
Sum	2.000	2.000	2.000	2.000	2.000	2.000	2.000	2.000	2.000	2.000	2.000
Al (VI)	0.024	0.037	0.004	0.003	0.022	0.027	0.057	0.021	0.005	0.041	0.040
Fe^{2+}	0.156	0.202	0.135	0.170	0.154	0.209	0.222	0.122	0.105	0.245	0.242
Fe^{3+}	0.130	0.067	0.126	0.088	0.041	0.092	0.063	0.058	0.079	0.059	0.069
Cr	0.000	0.000	0.002	0.000	0.020	0.001	0.002	0.016	0.022	0.000	0.000

(continued)

Table 2.2 (continued)

Sample	RU5					RU10B					
Suite/group	AkB-Tc					AkB-Tc					
Rock type	Tca					Tca					
Locality	Cóndores					Cóndores					
	p/c	p/r	mp/c	mp/r	gm	p/c	p/r	mp1/c	mp1/r	mp2/c	mp2/r
Mg	0.716	0.732	0.789	0.726	0.889	0.763	0.763	0.869	0.874	0.759	0.735
Mn	0.014	0.012	0.006	0.005	0.005	0.008	0.007	0.002	0.004	0.009	0.005
Ti	0.024	0.020	0.060	0.070	0.028	0.043	0.049	0.036	0.036	0.044	0.043
Ca	0.855	0.870	0.821	0.903	0.801	0.803	0.787	0.838	0.837	0.797	0.815
Na	0.081	0.060	0.056	0.036	0.041	0.054	0.049	0.038	0.038	0.046	0.051
Sum	2.000	2.000	1.999	2.001	2.001	2.000	2.000	2.000	2.000	2.000	2.000
Ca	45.70	46.20	43.74	47.73	42.38	42.83	42.73	44.36	44.08	42.64	43.68
Mg	38.27	38.87	42.04	38.37	47.04	40.69	41.42	46.00	46.02	40.61	39.39
Fe*	16.03	14.92	14.22	13.90	10.58	16.48	15.85	9.63	9.90	16.75	16.93

Sample	PM4					PM1					PM6		
Suite/group	AkB-Tc					AkB-Tc					AkB-Tc		
Rock type	AkB					Tca					Tc		
Locality	Almafuerte					Almafuerte					Almafuerte		
	p/c	p/r	mp/c	mp/r	gm	p/c	p/r	mp/c	mp/r	gm	p/c	p/r	gm
SiO$_2$	50.21	50.6	53.76	44.96	48.57	50.86	50.13	49.8	50.33	49.63	48.91	48.59	48.73
TiO$_2$	1.16	1.31	0.14	4.16	3.01	1.75	2.06	2.18	1.85	1.83	2.19	2.17	2.22
Al$_2$O$_3$	5.21	5.35	0.9	7.08	4.64	3.57	3.66	4.34	3.32	4.35	5.36	5.79	5.33
FeO$_{total}$	8.55	8.37	9.58	7.79	7.09	8.65	7.85	8.72	8.19	8.63	7.7	7.89	7.73
MnO	0.23	0.19	0.47	0.05	0.14	0.18	0.15	0.21	0.17	0.18	0.18	0.09	0.10
MgO	11.56	11.75	13.23	11.89	13.48	13.9	13.44	13.38	13.77	13.66	14.28	13.91	14.20
CaO	20.88	20.59	21.64	22.65	23.79	21.39	22.6	22.06	22.56	21.64	21.67	22.15	21.84
Na$_2$O	1.86	2.02	0.77	0.62	0.47	0.69	0.53	0.72	0.50	0.72	0.70	0.67	0.61
Cr$_2$O$_3$	0.02	0.01	0.00	0.00	0.00	0.04	0.00	0.00	0.00	0.05	0.20	0.12	0.08
Sum	99.68	100.19	100.49	99.20	101.19	101.03	100.41	101.43	100.69	100.69	101.2	101.38	100.85
Fe$_2$O$_3$*	4.09	3.94	0.6	3.79	2.47	2.02	1.73	3.21	2.53	3.56	4.31	4.66	4.12

	mp/r	gm
SiO$_2$	48.76	49.75
TiO$_2$	2.00	1.98
Al$_2$O$_3$	5.54	4.77
FeO$_{total}$	8.17	7.75
MnO	0.20	0.11
MgO	13.90	14.42
CaO	21.53	21.74
Na$_2$O	0.62	0.60
Cr$_2$O$_3$	0.03	0.09
Sum	100.75	101.22
Fe$_2$O$_3$*	3.89	3.27

(continued)

Table 2.2 (continued)

Sample	PM4					PM1					PM6				
Suite/group	AkB-Tc					AkB-Tc					AkB-Tc				
Rock type	AkB					Tca					Tc				
Locality	Almafuerte					Almafuerte					Almafuerte				
	p/c	p/r	mp/c	mp/r	gm	p/c	p/r	mp/c	mp/r	gm	p/c	p/r	mp/c	mp/r	gm
Si	1.863	1.865	1.996	1.694	1.789	1.870	1.858	1.827	1.859	1.83	1.787	1.774	1.787	1.792	1.818
Al (IV)	0.137	0.135	0.004	0.306	0.202	0.13	0.142	0.173	0.141	0.17	0.213	0.226	0.213	0.208	0.182
Sum	2.000	2.000	2.000	2.000	1.991	2.000	2.000	2.000	2.000	2.000	2.000	2.000	2.000	2.000	2.000
Al (VI)	0.091	0.098	0.035	0.008	0.000	0.025	0.018	0.015	0.004	0.019	0.018	0.023	0.017	0.032	0.023
Fe^{2+}	0.151	0.149	0.281	0.138	0.150	0.210	0.195	0.179	0.183	0.167	0.117	0.113	0.123	0.143	0.147
Fe^{3+}	0.114	0.109	0.017	0.107	0.069	0.056	0.048	0.089	0.07	0.099	0.118	0.128	0.114	0.108	0.090
Cr	0.001	0.000	0.000	0.000	0.000	0.001	0.000	0.000	0.000	0.001	0.006	0.004	0.002	0.001	0.003
Mg	0.639	0.645	0.732	0.668	0.740	0.762	0.743	0.732	0.758	0.751	0.778	0.757	0.776	0.762	0.786
Mn	0.007	0.006	0.015	0.002	0.005	0.006	0.005	0.007	0.005	0.006	0.006	0.003	0.003	0.006	0.003
Ti	0.032	0.036	0.004	0.118	0.083	0.048	0.057	0.06	0.051	0.051	0.06	0.06	0.061	0.055	0.055
Ca	0.830	0.813	0.861	0.914	0.939	0.843	0.897	0.867	0.893	0.855	0.848	0.866	0.858	0.848	0.851
Na	0.134	0.144	0.056	0.045	0.034	0.049	0.038	0.051	0.036	0.051	0.049	0.047	0.044	0.044	0.043
Sum	1.999	2.000	2.001	2.000	2.020	2.000	2.001	1.998	1.999	2.001	2.000	2.001	1.998	1.999	2.001
Ca	47.67	47.21	45.17	49.97	49.34	44.91	47.51	46.26	46.78	45.53	45.42	46.38	45.78	45.42	45.34
Mg	36.70	37.46	38.41	36.52	38.89	40.6	39.35	39.06	39.71	39.99	41.67	40.55	41.41	40.81	41.88
Fe*	15.62	15.33	16.42	13.50	11.77	14.49	13.14	14.67	13.51	14.48	12.91	13.07	12.81	13.77	12.79

Sample	Li1			Li4						
Suite/group	TrB-Lb			TrB-Lb						
Rock type	TrB			Lb						
Locality	Cóndores			Cóndores						
	p/c	mp/c	p/r	p/c	mp/r	gm	p/r	mp/c	mp/r	gm
SiO_2	51.10	53.08	50.98	50.60	50.29	51.99	49.97	50.75	50.71	48.89
TiO_2	0.65	1.30	0.75	1.51	1.91	1.24	1.63	0.73	1.77	2.06
Al_2O_3	4.57	2.01	5.18	4.13	2.93	2.83	4.45	3.05	3.53	4.27
FeO_{total}	7.38	4.46	7.36	6.59	6.53	3.95	6.78	9.08	5.96	7.00

(continued)

Table 2.2 (continued)

Sample	Li1					Li4				
Suite/group	TrB-Lb					TrB-Lb				
Rock type	TrB					Lb				
Locality	Cóndores					Cóndores				
	p/c	p/r	mp/c	mp/r	gm	p/c	p/r	mp/c	mp/r	gm
MnO	0.27	0.24	0.03	0.06	0.09	0.1	0.07	0.45	0.09	0.09
MgO	13.50	13.29	16.26	14.52	15.65	15.25	15.23	13.43	15.06	13.82
CaO	22.76	22.41	23.65	23.25	23.41	20.85	21.26	21.09	22.90	22.77
Na_2O	0.55	0.63	0.24	0.32	0.31	0.71	0.87	0.90	0.46	0.49
Cr_2O_3	0.04	0.04	0.44	0.08	1.23	0.15	0.23	0.00	0.57	0.08
Sum	100.83	100.87	101.47	99.89	100.70	99.89	100.49	99.46	101.05	99.47
$Fe_2O_3{}^*$	1.95	1.41	0.29	1.51	0.09	2.22	4.50	3.66	2.36	3.19
Si	1.875	1.870	1.920	1.866	1.897	1.861	1.825	1.894	1.851	1.821
Al (IV)	0.125	0.130	0.080	0.128	0.103	0.139	0.175	0.106	0.149	0.179
Sum	2.000	2.000	2.000	1.994	2.000	2.000	2.000	2.000	2.000	2.000
Al (VI)	0.073	0.094	0.006	0.000	0.019	0.040	0.016	0.028	0.003	0.008
Fe^{2+}	0.173	0.187	0.127	0.160	0.118	0.141	0.084	0.181	0.117	0.129
Fe^{3+}	0.054	0.039	0.008	0.042	0.003	0.062	0.124	0.103	0.065	0.089
Cr	0.001	0.001	0.013	0.002	0.036	0.004	0.007	0.000	0.016	0.002
Mg	0.739	0.727	0.877	0.803	0.851	0.836	0.829	0.747	0.819	0.767
Mn	0.009	0.008	0.001	0.002	0.003	0.003	0.002	0.014	0.003	0.003
Ti	0.018	0.021	0.035	0.053	0.034	0.042	0.045	0.020	0.049	0.058
Ca	0.895	0.880	0.916	0.924	0.915	0.822	0.832	0.843	0.895	0.908
Na	0.039	0.045	0.017	0.023	0.022	0.050	0.062	0.065	0.032	0.036
Sum	0.000	0.000	0.000	0.000	0.000	2.000	2.001	2.001	1.999	2.000
Ca	47.86	47.80	47.49	47.85	48.41	44.10	44.47	44.65	47.13	47.89
Mg	39.520	39.490	45.46	41.58	45.03	44.85	44.31	39.57	43.13	40.45
Fe*	12.620	12.710	7.05	10.56	6.56	11.05	11.22	15.78	9.74	11.66

Bsn basanite, *AKB* alkali basalt, *Tcb* trachybasalt, *Tca* trachyandesite, *TrB* transitional basalt, *Cóndores* Sierra de los Cóndores, *p* phenocryst, *mp* microphenocryst, *gm* microlite, *c* core, *r* rim
$Fe_2O_3{}^*$ calculated according Papike et al. (1974); $Fe^* = Fe^{2+} + Fe^{3+} + Mn$. Structural formulae on the basis of 4 cations and 6 oxygens

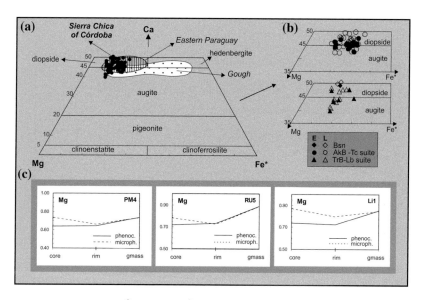

Fig. 2.10 a Ca–Mg–Fe* (Fe^{2+} + Mn + Fe^{3+}) plot for clinopyroxene of rocks of Sierra Chica of Córdoba, according to the classification of Morimoto (1988), in comparison with those from other alkaline provinces: eastern Paraguay (Cundari and Comin-Chiaramonti 1996) and Gough (le Roex 1985). **b** Ca–Mg–Fe* plots for clinopyroxenes of alkali basalt—trachyte suite (AkB-Tc) and transitional basalt—latibasalt suite (TrB-Lb) and the group of basanites (Bsn) shown in Lagorio (2008). *E* early clinopyroxenes (phenocrysts and microphenocryst cores), *L* late clinopyroxenes (phenocryst and microphenocrys rims and microlites of the groundmass). **c** Clinopyroxene diagrams showing Mg (a.f.u.) variation from core to rim of phenocryst and/or microphenocryst and the microlites of the groundmass (gmass)

Feldspar

Selected feldspar compositions of the SCC volcanic rocks are given in Table 2.4. Feldspars span from calcic andesine to sanidine in the An–Ab–Or diagram (Fig. 2.11). Macro-, pheno- and microphenocrysts of feldspars are characteristic of the salic rocks, while scarce or absent in the more basic ones. They correspond to plagioclase (An$_{54-40}$) ± anorthoclase (An$_{27-15}$Ab$_{45-63}$Or$_{28-22}$) and ± sanidine (An$_4$Ab$_{45}$Or$_{51}$). Groundmass of the different lithological types is generally characterized by the presence of plagioclase (An$_{52-27}$) and alkali feldspar (Or$_{67-15}$Ab$_{68-30}$An$_{27-1}$); nevertheless, in basanites, it is only composed of sodic anorthoclase (Or$_{11}$Ab$_{76}$An$_{13}$) as shown in Fig. 2.11.

Fe–Ti Oxides

Selected samples of homogeneous Fe–Ti oxides of the SCC volcanic rocks are presented in Table 2.5. They are common in the groundmass, scarce as

Table 2.3 Olivine microprobe compositions from selected rocks of Sierra Chica de Córdoba expanded from Lagorio (2008)

Sample	G3					EM5					CO7			
Suite/group	Bsn					AkB-Tc					AkB-Tc			
Rock type	Bsn					AkB					Tcb			
Locality	El Pungo					Cóndores					Cóndores			
	p/c	p/r	mp/c	mp/r	gm	p/c	p/r	mp/c	mp/r	gm	mp/c1	mp/r1	mp/c2	mp/r2
SiO_2	40.47	40.60	40.59	40.52	40.42	41.14	39.89	40.14	39.27	39.23	40.11	38.57	38.98	37.64
Al_2O_3	0.02	0.00	0.03	0.05	0.04	0.02	0.03	0.05	0.04	0.00	0.04	0.00	0.02	0.04
FeO_{total}	12.55	12.55	12.55	12.37	12.76	9.48	16.01	15.43	15.63	19.68	17.90	22.40	19.43	25.99
MnO	0.18	0.28	0.17	0.23	0.19	0.15	0.3	0.24	0.20	0.32	0.16	0.26	0.27	0.45
MgO	46.66	46.70	47.32	47.03	46.22	49.46	43.99	45.31	44.50	41.14	43.36	39.00	41.06	36.22
CaO	0.19	0.34	0.17	0.25	0.35	0.05	0.31	0.13	0.20	0.41	0.21	0.26	0.23	0.30
Sum	0.00	0.00	0.00	0.00	0.00	0.00	0.00	0.00	0.00	0.00	0.00	0.00	0.00	0.00
Si	1.003	1.002	0.998	1.000	1.004	1.002	1.001	0.996	0.992	0.999	1.000	0.998	0.999	0.992
Al (IV)	0.001	0.000	0.001	0.001	0.001	0.000	0.001	0.001	0.001	0.000	0.001	0.000	0.001	0.001
Sum	1.004	1.002	0.999	1.001	1.005	1.002	1.002	0.997	0.993	0.999	1.001	0.998	1.000	0.993
Fe^{2+}	0.260	0.259	0.258	0.255	0.265	0.193	0.336	0.320	0.330	0.419	0.373	0.485	0.417	0.573
Mn	0.004	0.006	0.004	0.005	0.004	0.003	0.006	0.005	0.004	0.007	0.003	0.006	0.006	0.010
Mg	1.724	1.719	1.735	1.730	1.712	1.797	1.646	1.676	1.675	1.562	1.612	1.504	1.569	1.422
Ca	0.005	0.009	0.005	0.007	0.009	0.001	0.008	0.003	0.005	0.011	0.006	0.007	0.006	0.009
Sum	1.993	1.993	2.002	1.997	1.990	1.994	1.996	2.004	2.014	1.999	1.994	2.002	1.998	2.014
Fo	86.72	86.64	86.89	86.92	86.42	90.16	82.78	83.75	83.36	78.56	81.06	75.42	78.78	70.94
Fa	13.09	13.06	12.93	12.83	13.38	9.69	16.90	16.00	16.43	21.09	18.77	24.3	20.92	28.56
Tph	0.19	0.29	0.18	0.25	0.20	0.16	0.32	0.25	0.21	0.35	0.17	0.28	0.30	0.50

Sample	EM3					MO1				
Suite/group	AkB-Tc					AkB-Tc				
Rock type	Tcb					AkB				
Locality	Cóndores					Los Molinos				
	p/c	p/r	mp/c	mp/r	gm	p/c	p/r	mp/c	mp/r	gm
SiO_2	40.39	41.08	37.97	37.99	37.90	40.69	40.61	40.1	40.12	40.25

(continued)

Table 2.3 (continued)

Sample	EM3					MO1				
Suite/group	AkB-Tc					AkB-Tc				
Rock type	Tcb					AkB				
Locality	Cóndores					Los Molinos				
	p/c	p/r	mp/c	mp/r	gm	p/c	p/r	mp/c	mp/r	gm
Al_2O_3	0.00	0.03	0.01	0.13	0.00	0.00	0.00	0.00	0.00	0.02
FeO_{total}	13.12	14.11	22.58	24.84	24.19	10.29	10.23	12.49	12.72	12.20
MnO	0.17	0.24	0.4	0.67	0.73	0.17	0.10	0.14	0.24	0.12
MgO	47.37	46.88	38.02	36.85	36.77	48.76	48.98	47.04	46.53	47.15
CaO	0.25	0.27	0.33	0.31	0.27	0.10	0.08	0.23	0.39	0.26
Sum	0.00	0.00	0.00	0.00	0.00	100.00	100.00	100.00	100.00	100.00
Si	0.992	1.000	0.998	0.994	0.998	0.999	0.996	0.995	0.997	0.996
Al (IV)	0.000	0.001	0.000	0.004	0.000	0.000	0.000	0.000	0.000	0.001
Sum	0.992	1.001	0.998	0.998	0.998	0.999	0.996	0.995	0.997	0.997
Fe^{2+}	0.269	0.287	0.496	0.544	0.533	0.211	0.210	0.259	0.264	0.252
Mn	0.003	0.005	0.009	0.015	0.016	0.003	0.002	0.003	0.005	0.003
Mg	1.734	1.700	1.489	1.438	1.444	1.784	1.792	1.740	1.724	1.739
Ca	0.006	0.007	0.009	0.009	0.008	0.002	0.002	0.006	0.010	0.007
Sum	2.012	1.999	2.003	2.006	2.001	2.000	2.006	2.008	2.003	2.001
Fo	86.41	85.34	74.67	72.02	72.45	89.26	89.42	86.91	86.48	87.21
Fa	13.42	14.41	24.88	27.23	26.74	10.57	10.48	12.95	13.26	12.66
Tph	0.17	0.25	0.45	0.74	0.81	0.17	0.10	0.14	0.25	0.13

Abbreviations as in Table 2.2. *Fo* forsterite, *Fa* fayalite, *Tph* tephroite

Table 2.4 Feldspar microprobe compositions from selected rocks of Sierra Chica de Córdoba expanded from Lagorio (2008)

Sample	G3	EM5				CO7			EM3		
Suite/Group	Bsn	AkB-Tc				AkB-Tc			AkB-Tc		
Rock type	Bsn	AkB				Tcb			Tcb		
Locality	Pungo	Cóndores				Cóndores			Cóndores		
	gm	gm1	gm2	gm3	gm4	gm1	gm2	gm3	gm1	gm2	gm3
SiO_2	64.56	62.94	60.79	62.58	58.92	63.93	64.56	56.83	56.85	64.32	63.07
TiO_2	0.05	0.21	0.25	0.29	5.61	0.22	0.28	0.21	0.20	0.19	0.22
Al_2O_3	21.49	22.00	23.90	22.30	16.83	20.2	18.94	26.68	26.88	20.08	21.17
FeO_{total}	0.35	0.46	0.35	0.31	5.13	0.33	0.30	0.61	0.61	0.27	0.27
CaO	2.66	3.46	5.53	3.79	0.13	1.75	0.56	8.76	8.76	1.56	2.74
Na_2O	8.94	7.96	7.70	8.08	3.89	5.69	3.56	6.27	6.27	4.08	5.51
K_2O	1.80	2.42	1.09	1.96	9.42	7.08	11.15	0.42	0.42	9.74	6.58
Sum	99.85	99.45	99.61	99.31	99.93	99.20	99.35	99.78	99.99	100.24	99.56
Or (wt%)	10.66	14.46	6.50	11.73	62.39	42.41	66.70	2.53	3.20	57.64	39.23
Ab (wt%)	76.05	68.17	65.78	69.23	36.89	48.82	30.49	53.59	52.33	34.62	47.04
An (wt%)	13.29	17.37	27.72	19.04	0.72	8.77	2.81	43.88	44.47	7.74	13.74

Sample	RU10B		PM4						PM1						
Suite/group	AkB-Tc		AkB-Tc						AkB-Tc						
Rock type	Tca		AkB						Tca						
Locality	Cóndores		Almafuerte						Almafuerte						
	Mp/c	Mp/r	p/c	p/r	mp/c	mp/r	gm1	gm2	gm1	gm2	p/c	p/r	mp/c	mp/r	gm
SiO_2	57.74	58.13	60.36	59.82	57.61	56.96	60.25	55.97	60.78	60.33	56.29	56.88	55.85	56.15	65.19
TiO_2	0.13	0.14	0.18	0.10	0.12	0.12	0.14	0.15	0.37	0.38	0.14	0.16	0.14	0.12	0.27

(continued)

Table 2.4 (continued)

Sample	RU10B								PM4		PM1				
Suite/group	AkB-Tc								AkB-Tc		AkB-Tc				
Rock type	Tca								AkB		Tca				
Locality	Cóndores								Almafuerte		Almafuerte				
	Mp/c	Mp/r	p/c	p/r	mp/c	mp/r	gm1	gm2	gm1	gm2	p/c	p/r	mp/c	mp/r	gm
Al_2O_3	26.07	26.16	23.65	24.00	26.26	27.06	23.62	27.36	23.53	24.76	27.28	26.42	26.99	27.45	19.54
FeO_{total}	0.42	0.47	0.29	0.30	0.44	0.49	0.39	0.54	0.32	0.39	0.46	0.33	0.40	0.40	0.27
CaO	8.04	8.02	5.43	5.84	8.22	9.03	5.43	9.52	5.23	6.35	9.38	8.53	9.23	9.54	0.91
Na_2O	5.50	5.67	5.33	5.58	5.71	5.38	5.60	5.23	6.88	6.65	5.34	5.67	5.27	5.43	6.07
K_2O	2.23	2.09	4.64	3.89	1.78	1.66	4.20	1.35	2.51	2.11	1.35	1.48	1.43	1.09	7.31
Sum	100.13	100.68	99.88	99.53	100.14	100.70	99.63	100.12	99.62	100.97	100.24	99.31	99.31	100.18	99.56
Or (wt%)	13.23	12.34	27.58	23.18	10.58	9.80	25.04	8.03	15.01	12.44	7.99	8.86	8.53	6.46	43.59
Ab (wt%)	46.74	47.92	45.33	47.60	48.49	45.46	47.79	44.50	58.78	56.13	45.35	48.42	45.11	46.04	51.86
An (wt%)	40.04	39.75	27.09	29.22	40.94	44.75	27.17	47.47	26.21	31.43	46.67	42.72	46.36	47.50	4.55

Sample	PM6							PM7					
Suite/group	AkB-Tc							AkB-Tc					
Rock type	Tc							Tcp					
Locality	Almafuerte							Almafuerte					
	Mp/c	Mp/r	p/c	p/r	mp/c	mp/r	gm	Mp/c	Mp/r	p/c	p/r	mp/c	mp/r
SiO_2	63.51	64.76	65.43	65.71	55.42	55.49	65.58	56.62	56.09	58.72	57.96	57.17	56.56
TiO_2	0.05	0.07	0.03	0.08	0.03	0.05	0.16	0.16	0.15	0.08	0.14	0.15	0.11
Al_2O_3	21.51	19.59	19.54	19.45	28.22	27.97	18.95	27.09	27.57	25.22	26.26	26.70	26.71
FeO_{total}	0.21	0.27	0.13	0.21	0.50	0.45	0.48	0.34	0.37	0.55	0.39	0.40	0.41

(continued)

Table 2.4 (continued)

Sample	PM6							PM7					
Suite/group	AkB-Tc							AkB-Tc					
Rock type	Tc							Tcp					
Locality	Almafuerte							Almafuerte					
	Mp/c	Mp/r	p/c	p/r	mp/c	mp/r	gm	Mp/c	Mp/r	p/c	p/r	mp/c	mp/r
CaO	2.92	1.05	0.85	0.71	10.35	10.13	0.33	9.14	9.66	7.10	8.14	8.69	8.84
Na$_2$O	7.36	5.54	5.50	5.12	5.13	5.21	4.25	5.36	5.28	6.34	5.88	5.62	5.56
K$_2$O	3.78	7.92	8.27	9.00	0.90	0.92	10.50	1.54	1.23	1.75	1.66	1.54	1.39
Sum	99.34	99.20	99.75	100.28	100.55	100.22	100.25	100.25	100.35	99.76	100.43	100.27	99.58
Or (wt%)	22.52	47.32	49.06	53.17	5.32	5.47	62.27	9.12	7.28	10.04	9.82	9.12	8.29
Ab (wt%)	62.85	47.40	46.71	43.30	43.37	44.17	36.08	45.44	44.73	54.07	49.79	47.66	47.46
An (wt%)	14.63	5.27	4.23	3.52	51.31	50.37	1.64	45.44	47.99	35.54	40.40	43.22	44.25

Sample	Li1			ES2					Li4			
Suite/group	TrB-Lb			TrB-Lb					TrB-Lb			
Rock type	TrB			TrB					Lb			
Locality	Cóndores			Cóndores					Cóndores			
	gm1	gm2	gm3	gm1	gm2	gm3	gm4	gm5	mp/c	mp/r	gm1	gm2
SiO$_2$	60.66	61.62	62.16	64.02	57.54	64.56	64.79	57.58	59.62	59.68	61.42	56.07
TiO$_2$	0.32	0.28	0.31	0.17	0.26	0.27	0.16	0.25	0.09	0.06	0.22	0.19
Al$_2$O$_3$	23.67	23.3	23.46	21.71	26.34	20.38	20.41	26.46	24.72	24.57	23.64	27.31
FeO$_{total}$	0.39	0.46	0.40	0.51	0.51	0.31	0.28	0.63	0.3	0.34	0.34	0.46
CaO	5.37	4.85	4.85	2.97	8.30	1.75	1.72	8.39	6.48	6.35	5.17	9.45
Na$_2$O	7.56	7.76	7.98	8.12	6.42	6.43	6.82	6.59	6.54	6.66	7.87	5.92

(continued)

Table 2.4 (continued)

Sample	Li1			ES2					Li4			
Suite/group	TrB-Lb			TrB-Lb					TrB-Lb			
Rock type	TrB			TrB					Lb			
Locality	Cóndores			Cóndores					Cóndores			
	gm1	gm2	gm3	gm1	gm2	gm3	gm4	gm5	mp/c	mp/r	gm1	gm2
K_2O	1.36	1.60	1.41	2.73	0.64	6.12	5.61	0.34	2.02	1.93	1.2	0.37
Sum	99.33	99.87	100.57	100.23	100.01	99.82	99.79	100.24	99.77	99.59	99.86	99.77
Or (wt%)	8.15	9.52	8.32	16.20	3.81	36.44	33.35	2.04	12.01	11.47	7.16	2.23
Ab (wt%)	64.83	66.23	67.59	69.01	54.71	54.83	58.09	56.07	55.65	56.78	67.00	50.51
An (wt%)	27.02	24.24	24.10	14.79	41.48	8.73	8.57	41.89	32.34	31.75	25.84	47.26

Abbreviations as in Table 2.2. *Or* orthoclase, *Ab* albite, *An* anorthite, *Mp* macrophenocryst

Fig. 2.11 Feldspar
compositions in relation to the
contents of Ab–An–Or,
according to the limits
proposed by Deer et al.
(1992). *Pl* plagioclase, *E* early
feldspar; *L* late feldspar.
Based on Lagorio (2008)

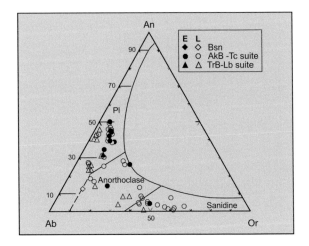

microphenocrysts and compositionally correspond to Ti-magnetite (Mt) and ilmenite (Ilm). Ti-magnetite is the common oxide phase, while ilmenite coexists in few
rocks.

Ti-magnetite compositions are mainly rich in ulvöspinel component (Ulv. = 85–
48 %), with the exception of a few samples (Ulv. = 14–4 %). The R_2O_3 component
of ilmenites ranges from 4 to 8 %.

Coexistence of Mt and Ilm yielded (Spencer and Lindsley 1981) temperatures
ranging between 1190 and 845 °C, and fO_2 values that straddle FQM buffer
(Fig. 2.12). Temperature decrease suggests different stages of post-eruptive
re-equilibration. This re-equilibration was accompanied by maghemitization and
haematite replacement, which both were interpreted as the cause for the acquisition
of a stable magnetic remanence in SCC volcanic rocks (Geuna et al. 2015).

2.1.1.6   Geochemistry

Previous chemical data of major elements have been presented by Gordillo and
Lencinas (1967a, b, 1969), as it was previously mentioned. Later, in the following
decades, several authors carried out geochemical studies in different areas of Sierra
Chica, as Cortelezzi et al. (1981), Kay and Ramos (1996), Sánchez and Bermúdez
(1997), Lagorio et al. (1997), Lagorio (1998, 2003, 2008), Escayola et al. (1998,
1999), Lucassen et al. (2002) and Ancheta et al. (2002), most of them including also
the study of trace elements. Major and trace element contents provided by Lagorio
(2003, 2008) were determined at the Dipartimento di Scienze della Terra,
University of Trieste, by using a PW-1404 XRF spectrometer and the procedures of
Philips (1994) for the correction of matrix effects. Major element abundance was

Table 2.5 Magnetite and ilmenite microprobe compositions from selected rocks of Sierra Chica de Córdoba expanded from Lagorio (2008)

Sample	G3	MO1	EM5	CO7	EM3	RU5	Li8	RU10B	PM4	PM1	PM6	PM7	Li1	ES2	Li4	CN1
Suite/group	Bsn	AkB-Tc	AkB-Tc	AkB-Tc	AkB-Tc	AkB-Tc	AkB-Tc	AkB-Tc	AkB-Tc	AkB-Tc	AkB-Tc	AkB-Tc	TrB-Lb	TrB-Lb	TrB-Lb	TrB-Lb
Rock type	Bsn	AkB	AkB	Tcb	Tcb	Tca	Tca	Tca	AkB	Tca	Tc	Tcp	TrB	TrB	TrB	TrB
Locality	Pungo	Molinos	Cónd.	Cónd.	Cónd.	Cónd.	Cónd.	Cónd.	Almaf.	Almaf.	Almaf.	Almaf.	Cónd.	Cónd.	Cónd.	Cónd.
	gm	gm	gm	gm	gm	gm	gm	gm	gm	gm	gm	gm	gm	gm	gm	gm
Magnetite																
SiO_2	0.23	0.13	0.04	0.26	0.27	0.24	0.27	0.24	0.27	0.11	0.14	0.28	0.28	0.24	0.05	0.28
TiO_2	30.00	17.06	22.21	17.70	20.87	1.22	23.11	4.82	29.99	29.36	18.28	27.97	30.25	2.40	21.28	25.05
Al_2O_3	0.31	2.86	2.52	0.68	1.67	4.00	0.28	3.54	0.34	1.33	1.13	1.45	0.72	0.94	1.45	0.47
FeO_{total}	65.45	74.35	71.41	77.77	74.32	83.86	73.47	82.99	65.45	65.24	75.71	66.26	65.46	90.3	72.69	70.91
MnO	1.68	0.79	1.10	2.21	1.02	1.17	0.29	0.43	1.35	2.13	0.85	1.45	0.68	0.10	0.38	0.29
MgO	0.08	2.28	0.60	0.08	0.17	3.61	0.29	1.63	0.28	0.03	1.11	0.92	1.37	1.00	2.11	1.77
CaO	0.29	0.00	0.10	0.11	0.41	1.10	0.35	1.85	1.03	0.09	0.07	0.27	0.36	0.12	0.04	0.55
Cr_2O_3	0.05	0.06	0.21	0.00	0.01	0.00	0.15	0.00	0.08	0.13	0.02	0.00	0.07	0.00	0.49	0.00
Total	98.09	97.53	98.19	98.81	98.74	95.20	98.21	95.50	98.79	98.42	97.31	98.60	99.19	95.10	98.49	99.32
FeO	56.84	43.56	50.21	46.20	49.75	25.67	51.97	31.69	56.27	56.23	45.95	54.44	56.50	32.45	47.86	51.56
Fe_2O_3	9.57	34.21	23.55	35.07	27.29	64.66	23.89	57.00	10.20	10.01	33.06	13.14	9.96	64.28	27.59	21.50
Total	99.06	100.95	100.55	102.31	101.48	101.66	100.6	101.19	99.82	99.42	100.60	99.91	100.20	101.52	101.26	101.47
% Ulv.	85.92	48.03	62.82	50.19	59.35	4.28	65.88	14.48	85.10	83.69	51.85	79.43	85.05	7.70	59.17	69.78
Ilmenite																
SiO_2			0.24	0.01	0.06		0.15			0.03						
TiO_2			49.64	49.41	50.95		48.41			50.28						
Al_2O_3			0.79	0.00	0.04		0.2			0.25						
FeO_{total}			43.32	45.57	42.82		48.32			41.68						
MnO			0.87	0.70	0.83		0.23			2.52						
MgO			3.47	2.11	3.32		1.07			3.99						
CaO			0.11	0.06	0.14		0.20			0.17						
Cr_2O_3			0.06	0.00	0.03		0.23			0.01						

(continued)

Table 2.5 (continued)

Sample	G3	MO1	EM5	CO7	EM3	RU5	Li8	RU10B	PM4	PM1	PM6	PM7	Li1	ES2	Li4	CN1
Suite/group	Bsn	AkB-Tc	AkB-Tc	AkB-Tc	AkB-Tc	AkB-Tc	AkB-Tc	AkB-Tc	AkB-Tc	AkB-Tc	AkB-Tc	AkB-Tc	TrB-Lb	TrB-Lb	TrB-Lb	TrB-Lb
Rock type	Bsn	AkB	AkB	Tcb	Tcb	Tca	Tca	Tca	AkB	Tca	Tc	Tcp	TrB	TrB	TrB	TrB
Locality	Pungo	Molinos	Cónd.	Cónd.	Cónd.	Cónd.	Cónd.	Cónd.	Almaf.	Almaf.	Almaf.	Almaf.	Cónd.	Cónd.	Cónd.	Cónd.
	gm	gm	gm	gm	gm	gm	gm	gm	gm	gm	gm	gm	gm	gm	gm	gm
Total			0.00	0.00	98.19		0.00			0.00						
FeO			37.73	39.89	38.94		41.31			35.37						
Fe$_2$O$_3$			6.21	6.31	4.31		7.79			7.01						
Total			99.12	98.49	98.63		99.58			99.63						
% R$_2$O$_3$			7.04	6.00	4.14		7.93			6.88						
T (°C)			980	845	848		1040			1190						
log fO$_2$			−11.5	−14.1	−14.2		−10.4			−8.7						

Abbreviations as in Table 2.2. FeO, Fe$_2$O$_3$, R$_2$O$_3$, and Ulvöspinel (Ulv.) calculated according to Carmichael (1967). Temperature and log fO$_2$ following Spencer and Lindsley (1981) for homogeneous magnetite–ilmenite pairs. *Cónd.* Sierra de los Cóndores. *Almaf.* Almafuerte

Fig. 2.12 Temperature (°C) versus $-\log fO_2$ (Spencer and Lindsley 1981) for homogeneous groundmass magnetite–ilmenite pairs of Sierra Chica. *HM* haematite–magnetite; *NNO* nickel–nickel oxide; *FQM* fayalite–quartz–magnetite; *MW* magnetite–wüstite. Based on Lagorio (2008)

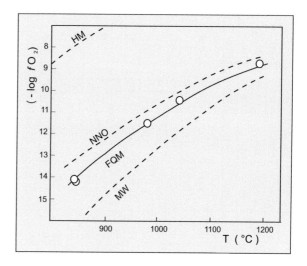

recalculated to 100 % on a volatile-free basis; FeO was obtained by titration and loss on ignition (L.O.I.) corrected for FeO oxidation, gravimetrically. The analytical uncertainties are estimated at less than 5 and 10 % for major and trace elements, respectively. Rare earth elements (REE), Th, Ta, Hf, U and Ga of selected rocks were measured by inductively coupled mass spectrometry (ICP-MS) at Activation Laboratory (Canada).

Representative analyses are given in Table 2.6, whereas the complete set of data is in Lagorio (2003) and available on request. CIPW-normative parameters and mg# (at. $Mg/(Mg + Fe^{2+})$) values were calculated assuming $Fe_2O_3/FeO = 0.18$.

As it has been previously pointed out, major element composition allowed the distinction of the rocks in several groups: (1) alkali basalt—trachyte suite, (2) transitional basalt—latibasalt suite, (3) basanites and (4) ankaratrites (Fig. 2.8b). In these graphics, the samples presented by other authors whose compositions are complementary to those obtained by Lagorio (2003, 2008) have been included. It should be noted that a suite consisting of basanite–phonolite is depicted, as was defined by Ferreira Pittau et al. (2008) considering several samples from the southern sector of Sierra de los Cóndores provided by other authors.

Major element variation diagrams of samples presented by Lagorio (2003, 2008) show a strong overlap between alkaline and transitional suites, with slightly higher values of SiO_2 and subtle lower of CaO and P_2O_5 in the transitional suite (Fig. 2.13). Regarding trace elements, this suite has the lowest values of Nb (Fig. 2.14). Basanites display a slightly lower content of SiO_2 and TiO_2, as well as slightly higher FeO_t and CaO tenors, for the same MgO values regarding both alkaline and transitional suites. They are distinguished by having the lowest Zr content of all studied rocks, low tenors of LREE and high Nb and Sr values (Fig. 2.14). Ankaratrites are characterized by displaying the lowest percentages of SiO_2 and Al_2O_3, high contents of TiO_2 and the highest CaO, P_2O_5, Ba and LREE values (Figs. 2.13 and 2.14).

Table 2.6 Major and trace element compositions of representative volcanic rocks of Sierra Chica of Córdoba expanded from Lagorio (2008)

Sample	G3	LM3	G2	G1	MO1	MO19	EM4	EM5	RU13	CN7	CN6	EM3	CO7
Suite/group	Bsn	Bsn	Bsn	Bsn	AkB	AkB-Tc	AkB-Tc	AkB-Tc	AkB-Tc	AkB-Tc	AkB-Tc	AkB-Tc	AkB-Tc
Rock type	Bsn	Bsn	Bsn	Bsn	AkB	AkB	AkB	AkB	AkB	AkB	AkB	Tcb	Tcb
Locality	Pungo	Pungo	Pungo	Pungo	Molinos	Molinos	Cónd.	Cónd.	Cónd.	Cónd.	Cónd.	Cónd.	Cónd..
SiO_2 (wt%)	45.07	44.87	45.84	46.05	47.94	46.92	46.82	46.98	50.80	49.05	49.77	51.08	51.46
TiO_2	2.68	2.71	2.60	2.61	3.15	3.68	2.90	2.89	2.22	3.47	3.42	2.32	2.91
Al_2O_3	9.92	10.06	10.48	10.82	10.36	10.94	9.71	9.84	13.53	11.03	11.61	14.21	13.19
FeO_{total}	11.09	11.24	10.75	10.52	9.81	11.27	10.61	10.60	9.12	10.98	10.49	8.78	8.82
MnO	0.16	0.17	0.16	0.16	0.16	0.17	0.16	0.16	0.13	0.17	0.16	0.14	0.12
MgO	15.51	14.61	13.85	13.30	13.20	11.81	15.30	14.89	10.14	9.68	8.28	7.61	8.31
CaO	10.79	10.68	10.86	10.77	9.81	10.07	9.89	9.77	7.97	10.24	9.95	8.05	6.98
Na_2O	2.52	3.68	3.15	3.28	2.28	3.06	2.73	2.96	3.90	2.87	2.99	4.06	3.91
K_2O	1.57	1.41	1.57	1.72	2.59	1.28	1.13	1.18	1.55	1.91	2.57	3.23	3.43
P_2O_5	0.70	0.58	0.75	0.78	0.70	0.80	0.75	0.75	0.56	0.60	0.76	0.52	0.88
mg-no.	0.74	0.73	0.73	0.72	0.74	0.68	0.75	0.74	0.70	0.65	0.62	0.64	0.66
FeO (wt%)	5.36	5.30	6.08	6.42	n.m.	n.m.	6.25	6.58	n.m.	2.66	2.31	5.74	5.78
Fe_2O_3 (wt%)	6.33	6.57	5.16	4.52	n.m.	n.m.	4.81	4.43	n.m.	9.21	9.07	3.36	3.36
L.O.I. (wt%)	4.96	3.63	4.02	4.21	5.30	4.80	4.67	4.50	5.41	5.06	5.49	2.42	3.53
Q (CIPW)													
Ne (CIPW)	8.20	14.48	9.91	10.40	3.08	3.33	3.25	4.18	0.99	1.18	1.65	6.71	3.23
Hy (CIPW)													
Ol (CIPW)	25.93	23.19	22.34	21.25	21.04	19.81	26.03	25.24	19.39	15.09	12.52	13.13	15.10
Cr (ppm)	591	627	630	624	317	308	653	632	284	411	325	282	205
Ni	443	430	423	397	249	266	464	468	235	289	249	152	167
Rb	53	21	48	55	60	22	14	16	6	44	55	30	30
Ba	970	993	999	992	1071	1181	918	921	789	1077	963	745	702
Sr	1229	1139	1174	1207	2318	1840	994	903	961	1107	1120	1100	936
Nb	87	86	85	89	82	95	83	78	78	83	86	73	80

(continued)

Table 2.6 (continued)

Sample	G3	LM3	G2	G1	MO1	MO19	EM4	EM5	RU13	CN7	CN6	EM3	CO7
Suite/group	Bsn	Bsn	Bsn	Bsn	AkB-Tc	AkB-Tc	AkB-Tc	AkB-Tc	AkB-Tc	AkB-Tc	AkB-Tc	AkB-Tc	AkB-Tc
Rock type	Bsn	Bsn	Bsn	Bsn	AkB	AkB	AkB	AkB	AkB	AkB	AkB	Tcb	Tcb
Locality	Pungo	Pungo	Pungo	Pungo	Molinos	Molinos	Cónd.	Cónd.	Cónd.	Cónd.	Cónd.	Cónd.	Cónd.
Zr	194	198	204	195	409	468	287	298	422	313	399	376	544
Y	21	21	22	21	27	31	26	25	23	30	29	23	25
La	58.5	58.2	57	58	98.8	91.3	65.3	68	73.3	69	62.7	71.9	74.4
Ce	109.0	110.0	100	106	166.0	164.0	121.0	125	128.0	118	121.0	135.1	148.0
Pr	11.49	11.84			16.97	17.39	13.66		14.9		14.06	14.6	16.01
Nd	49.9	49.9	50	49	66.2	66.6	56.1	53	56.5	56	60.7	53.9	65.6
Sm	9.93	10.10			12.20	13.20	10.90		9.52		12.10	8.87	11.60
Eu	3.00	3.05			3.29	3.89	3.30		2.96		3.74	2.91	3.34
Gd	7.75	7.78			8.50	10.10	8.42		8.12		9.45	7.09	8.43
Tb	1.10	1.14			1.23	1.45	1.24		1.11		1.37	1.01	1.16
Dy	5.24	5.32			5.96	7.03	6.20		5.35		6.58	4.96	5.45
Ho	0.87	0.89			0.95	1.17	1.02		0.91		1.07	0.85	0.94
Er	2.13	2.24			2.49	3.08	2.71		2.44		2.75	2.23	2.39
Tm	0.23	0.24			0.31	0.37	0.31		0.30		0.31	0.27	0.26
Yb	1.37	1.42			1.70	1.89	1.78		1.83		1.84	1.61	1.54
Lu	0.17	0.18			0.25	0.29	0.25		0.25		0.24	0.22	0.31
Hf	4.9	5.0			8.0	9.7	6.4		8.6		8.3	8.0	12.0
Ta	4.11	4.19			4.95	5.26	3.87		4.20		4.39	3.90	4.82
Th	7.82	7.36			10.30	9.57	6.60		8.04		11.40	9.03	7.99
U	1.66	1.36			1.91	1.80	1.12		1.43		1.33	1.72	1.71
Ga	19	19			20	22	19		25		22	21	19
(La/Yb)$_{cn}$	28.79	27.63			39.18	32.57	24.73		27.00		22.79	30.11	32.31
(Eu/Eu*)$_{cn}$	1.05	1.05			0.99	1.03	1.06		1.03		1.07	1.12	1.03

(continued)

Table 2.6 (continued)

Sample	CO12	RU5	RU7	Li8	RU10B	RU12B	PM4	PM2	PM1	PM5	PM7	PM6	PM6B
Suite/Group	AkB-Tc	AkB-Tc	AkB-Tc	AkB-Tc	AkB-Tc	AkB-Tc	AkB-Tc	AkB-Tc	AkB-Tc	AkB-Tc	AkB-Tc	AkB-Tc	AkB-Tc
Rock type	Tcb	Tca	Tca	Tca	Tca	Tca	AkB	Tcb	Tca	Tca	Tcp	Tc	QTc
Locality	Cónd.	Cónd.	Cónd.	Cónd.	Cónd.	Cónd.	Almaf.	Almaf.	Almaf.	Almaf.	Almaf.	Almaf.	Almaf.
Sample	CO12	RU5	RU7	Li8	RU10B	RU12B	PM4	PM2	PM1	PM5	PM7	PM6	PM6B
Suite/Group	AkB-Tc	AkB-Tc	AkB-Tc	AkB-Tc	AkB-Tc	AkB-Tc	AkB-Tc	AkB-Tc	AkB-Tc	AkB-Tc	AkB-Tc	AkB-Tc	AkB-Tc
Rock type	Tcb	Tca	Tca	Tca	Tca	Tca	AkB	Tcb	Tca	Tca	Tcp	Tc	QTc
Locality	Cónd.	Cónd.	Cónd.	Cónd.	Cónd.	Cónd.	Almaf.	Almaf.	Almaf.	Almaf.	Almaf.	Almaf.	Almaf.
SiO_2 (wt%)	50.08	56.31	56.07	56.04	56.69	56.72	47.67	52.14	52.36	58.94	55.28	62.62	63.15
TiO_2	3.10	2.04	2.15	2.18	1.78	1.76	3.12	2.66	2.59	1.57	2.05	0.94	1.04
Al_2O_3	12.95	15.09	14.97	14.81	17.38	17.48	11.13	15.14	15.47	16.14	17.59	17.55	17.31
FeO_{total}	9.31	6.51	6.74	6.92	6.51	6.25	11.03	8.62	8.30	5.58	6.49	3.67	4.01
MnO	0.13	0.10	0.13	0.11	0.09	0.10	0.16	0.12	0.11	0.12	0.10	0.10	0.08
MgO	8.48	5.58	5.13	5.01	3.12	2.89	11.63	5.96	5.95	3.38	3.12	1.12	1.47
CaO	8.02	4.92	5.31	5.21	4.59	4.62	9.27	7.67	6.20	4.53	4.76	1.46	1.32
Na_2O	3.33	3.70	3.77	3.64	3.77	4.63	3.31	3.35	4.70	4.01	6.13	3.65	2.83
K_2O	3.71	5.13	5.11	5.52	5.30	4.77	1.88	3.65	3.59	5.33	3.85	8.62	8.52
P_2O_5	0.89	0.61	0.61	0.56	0.77	0.78	0.80	0.69	0.72	0.41	0.62	0.25	0.27
mg-no.	0.65	0.64	0.61	0.60	0.50	0.49	0.69	0.59	0.60	0.56	0.50	0.39	0.43
FeO (wt%)	6.56	1.24	1.9	1.23	0.99	0.80	2.78	1.45	1.71	1.36	1.98	1.61	0.41
Fe_2O_3 (wt%)	3.03	5.84	5.37	6.31	6.12	6.04	9.14	7.94	7.31	4.68	5.00	2.28	3.99
L.O.I. (wt%)	3.15	2.75	2.95	3.11	3.43	2.60	4.89	4.99	3.71	2.92	2.66	1.33	2.60
Q (CIPW)								0.23	5.37	1.08	7.30	2.47	7.88
Ne (CIPW)	4.49	6.98	3.95	7.28	12.31	2.90	5.03						
Hy (CIPW)								11.66	12.10	10.35	7.15	6.80	7.91
Ol (CIPW)	14.60	6.43	7.24	8.53	0.70	6.32	20.33						
Cr (ppm)	227	89	88	99	20	12	357	107	81	50	20	16	26
Ni	173	77	70	65	14	12	311	78	73	33	10	5	12
Rb	46	94	92	109	139	107	42	49	51	143	92	225	199
Ba	740	825	738	765	1155	1494	1213	972	990	868	1109	736	684
Sr	1152	871	863	762	962	973	919	797	846	601	949	224	192

(continued)

Table 2.6 (continued)

Sample	CO12	RU5	RU7	Li8	RU10B	RU12B	PM4	PM2	PM1	PM5	PM7	PM6	PM6B
Suite/Group	AkB-Tc	AkB-Tc	AkB-Tc	AkB-Tc	AkB-Tc	AkB-Tc	AkB-Tc	AkB-Tc	AkB-Tc	AkB-Tc	AkB-Tc	AkB-Tc	AkB-Tc
Rock type	Tcb	Tca	Tca	Tca	Tca	Tca	AkB	Tcb	Tca	Tca	Tcp	Tc	QTc
Locality	Cónd.	Cónd.	Cónd.	Cónd.	Cónd.	Cónd.	Almaf.	Almaf.	Almaf.	Almaf.	Almaf.	Almaf.	Almaf.
Nb	86	84	86	88	94	93	91	91	91	82	86	111	111
Zr	510	667	670	715	649	645	387	536	543	568	539	743	775
Y	27	27	28	30	35	35	26	28	27	34	28	43	44
La	80	100	84.0	91.9	95.0	107	70.3	87	80.6	81.1	89.4	98.1	94
Ce	157	176	161.0	169.3	181.0	188	130.0	157	144.7	146.6	158.0	178.0	171
Pr			15.94	17.80	18.04		14.05		15.30	15.30	16.05	17.47	
Nd	65	66	60.3	62.4	67.6	74	58.5	61	56.3	54.0	61.3	61.4	62
Sm			10.10	9.77	11.40		10.60		8.98	8.95	10.00	10.50	
Eu			2.74	2.94	2.94		3.26		2.75	2.63	2.98	2.14	
Gd			7.31	7.47	8.03		8.44		7.24	7.20	7.04	7.44	
Tb			1.05	1.04	1.22		1.19		1.05	1.09	1.10	1.25	
Dy			5.06	5.24	5.90		5.93		5.19	5.71	5.58	6.40	
Ho			0.89	0.89	1.06		0.98		0.89	1.03	0.96	1.19	
Er			2.37	2.39	2.93		2.50		2.41	2.93	2.72	3.50	
Tm			0.29	0.30	0.36		0.29		0.31	0.39	0.33	0.50	
Yb			1.72	1.82	2.26		1.57		1.78	2.39	1.99	3.15	
Lu			0.24	0.25	0.31		0.22		0.25	0.33	0.29	0.44	
Hf			13.0	12.6	12.0		8.0		9.4	10.5	11.0	14.0	
Ta			4.54	4.70	4.82		4.30		5.00	4.60	4.77	6.46	
Th			14.50	15.00	19.50		9.05		13.20	15.50	14.80	24.70	
U			1.75	1.92	2.07		1.70		2.20	2.49	2.84	3.08	
Ga			25	26	28		21		25	22	28	26	
$(La/Yb)_{cn}$			32.93	33.77	28.34		29.94		30.28	22.69	30.04	20.83	
$(Eu/Eu^*)_{cn}$			0.98	1.05	0.94		1.05		1.04	1.00	1.09	0.74	

(continued)

Table 2.6 (continued)

	Li1	ES2	CO18	CN14	CO8	Li4	CN4	CN1	PM3	D4	D7	MO5	MO6
Sample	Li1	ES2	CO18	CN14	CO8	Li4	CN4	CN1	PM3	D4	D7	MO5	MO6
Suite/Group	TrB-Lb	TrB-Lb	TrB-Lb	TrB-Lb	TrB-Lb	TrB-Lb	TrB-Lb	TrB-Lb	TrB-Lb	TrB-Lb	TrB-Lb	Akt	Akt
Rock type	TrB	TrB	TrB	TrB	Lb	Lb	Lb	Lb	Lb	TrB	TrB	Akt	Akt
Locality	Cónd.	Cónd.	Cónd.	Cónd.	Cónd.	Cónd.	Cónd.	Cónd.	Almaf.	Despeñ.	Despeñ.	Molinos	Molinos
Sample	Li1	ES2	CO18	CN14	CO8	Li4	CN4	CN1	PM3	D4	D7	MO5	MO6
Suite/Group	TrB-Lb	TrB-Lb	TrB-Lb	TrB-Lb	TrB-Lb	TrB-Lb	TrB-Lb	TrB-Lb	TrB-Lb	TrB-Lb	TrB-Lb	Akt	Akt
Rock type	TrB	TrB	TrB	TrB	Lb	Lb	Lb	Lb	Lb	TrB	TrB	Akt	Akt
Locality	Cónd.	Cónd.	Cónd.	Cónd.	Cónd.	Cónd.	Cónd.	Cónd.	Almaf.	Despeñ.	Despeñ.	Molinos	Molinos
SiO_2 (wt%)	48.57	51.03	49.34	49.87	53.85	52.70	53.11	55.00	53.01	52.31	52.71	43.11	43.09
TiO_2	2.77	2.89	3.97	2.76	2.39	2.31	2.46	2.12	2.70	2.51	2.45	3.80	4.27
Al_2O_3	10.38	11.47	10.73	11.22	14.23	13.45	12.69	13.85	15.24	11.03	11.82	8.56	8.72
FeO_{total}	9.85	9.05	10.76	9.70	7.34	7.91	8.99	8.14	8.43	9.50	9.19	11.12	10.69
MnO	0.15	0.14	0.17	0.14	0.09	0.14	0.13	0.12	0.12	0.10	0.11	0.19	0.17
MgO	13.87	12.50	12.37	12.35	9.18	8.98	8.19	7.08	5.91	11.13	10.02	13.45	10.88
CaO	9.60	7.63	7.49	8.22	5.42	7.25	7.81	6.30	7.00	7.96	7.84	14.44	16.51
Na_2O	2.23	2.32	2.18	2.76	3.41	2.91	3.13	3.90	3.08	2.62	2.97	3.36	2.79
K_2O	1.96	2.34	2.23	2.35	3.40	3.60	2.99	2.81	3.81	2.39	2.46	1.05	1.75
P_2O_5	0.62	0.63	0.75	0.63	0.68	0.75	0.52	0.67	0.69	0.46	0.42	0.92	1.13
mg-no.	0.74	0.74	0.70	0.73	0.72	0.70	0.65	0.64	0.59	0.71	0.69	0.71	0.68
FeO (wt%)	2.00	2.34	2.40	3.5	1.61	1.64	2.06	1.9	1.73	1.21	1.84	n.m.	n.m.
Fe_2O_3 (wt%)	8.7	7.43	9.27	6.87	6.36	6.94	7.68	6.92	7.43	9.19	8.14	n.m.	n.m.
L.O.I. (wt%)	5.03	2.21	4.61	4.33	4.64	4.67	4.16	3.24	5.27	7.78	5.75	10.56	9.80
Q (CIPW)													
Ne (CIPW)													
Hy (CIPW)	0.06	17.16	17.45	1.07	13.72	7.28	9.15	15.75	9.90	15.49	12.23		
Ol (CIPW)	23.29	10.79	11.00	21.36	9.03	11.56	8.15	3.32	5.38	8.85	9.18		
Cr (ppm)	492	322	349	457	176	270	286	189	104	446	433	553	417
Ni	366	250	291	314	157	151	198	178	81	298	316	325	301
Rb	11	37	43	48	41	28	72	61	50	61	60	13	18
Ba	1057	764	827	915	728	852	885	1083	951	665	668	1528	1677
Sr	974	1050	923	886	997	978	1017	1064	726	947	767	754	728

(continued)

Table 2.6 (continued)

Sample	Li1	ES2	CO18	CN14	CO8	Li4	CN4	CN1	PM3	D4	D7	MO5	MO6
Suite/Group	TrB-Lb	TrB-Lb	TrB-Lb	TrB-Lb	TrB-Lb	TrB-Lb	TrB-Lb	TrB-Lb	TrB-Lb	TrB-Lb	TrB-Lb	Akt	Akt
Rock type	TrB	TrB	TrB	TrB	Lb	Lb	Lb	Lb	Lb	TrB	TrB	Akt	Akt
Locality	Cónd.	Cónd.	Cónd.	Cónd.	Cónd.	Cónd.	Cónd.	Cónd.	Almaf.	Despeñ.	Despeñ.	Molinos	Molinos
Nb	79	60	65	69	72	88	59	52	90	60	52	92	99
Zr	436	253	432	279	484	538	271	227	508	251	244	609	577
Y	21	25	43	22	22	25	26	26	27	23	25	28	30
La	91.6	63.9	69	71.4	77.6	93.4	64	76.8	83	52	45	120.0	144.0
Ce	165.0	123.0	134	134.0	147.0	174.0	118	134.0	151	100	98	214.0	245.0
Pr	16.63	13.45		14.90	16.00	17.58		15.12				22.01	26.22
Nd	62.6	56.8	67	59.9	60.3	66.0	60	61.8	57	47	47	79.6	98.6
Sm	10.10	11.40		11.00	9.85	11.20		11.40				13.30	16.10
Eu	2.96	3.32		3.30	3.12	3.10		3.41				3.65	4.11
Gd	7.43	8.51		8.06	7.62	7.82		8.40				10.50	12.20
Tb	1.05	1.26		1.20	1.00	1.16		1.20				1.37	1.60
Dy	5.15	5.90		5.84	4.96	5.49		5.73				6.22	7.28
Ho	0.87	1.02		0.95	0.81	0.95		0.91				1.01	1.12
Er	2.37	2.58		2.55	2.07	2.58		2.38				2.74	3.15
Tm	0.27	0.28		0.28	0.24	0.29		0.27				0.31	0.36
Yb	1.51	1.70		1.64	1.41	1.81		1.49				1.53	1.80
Lu	0.22	0.22		0.22	0.18	0.25		0.21				0.23	0.26
Hf	9.6	6.3		6.5	9.2	11.0		6.0				13.0	14.0
Ta	4.06	3.09		3.46	3.80	4.82		2.34				5.40	6.28
Th	8.76	8.37		7.14	8.93	12.10		9.24				10.10	11.20
U	0.96	1.29		1.29	1.52	2.02		0.84				1.97	1.87
Ga	20	22		21	23	24	24	24				19	20
$(La/Yb)_{cn}$	40.56	25.13		29.11	36.80	34.51		34.47				52.88	53.94
$(Eu/Eu*)_{cn}$	1.04	1.03		1.07	1.10	1.01		1.07				0.94	0.90

Major elements recalculated to 100 % on a volatile-free basis. mg-no = $Mg/(Mg + Fe^{2+})$ and CIPW-normative compositions assuming $Fe_2O_3/FeO = 0.18$. Q, Ne, Hy, and Ol = CIPW-normative quartz, nepheline, hypersthene, and olivine, respectively. Trace element contents in normal and italic styles indicate measurement by XRF and ICP-MS techniques, respectively. *Bsn* basanite, *AkB* alkali basalt, *Tcb* trachybasalt, *Tca* trachyandesite, *Tc* trachyte, *QTc* quartz trachyte, *Tcp* quartz trachyphonolite, *TrB* transitional basalt, *Lb* latibasalt; *Pungo* El Pungo; *Molinos* Los Molinos; *Cónd.* Sierra de los Cóndores; *Almaf.* Almafuerte; *Despeñ.* Despeñaderos; *cn* chondrite normalized (Boynton 1984).

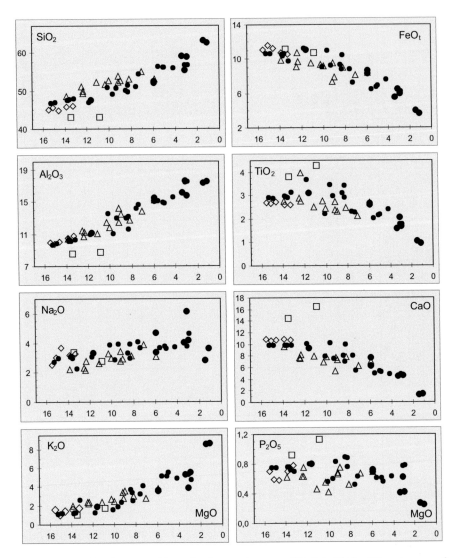

Fig. 2.13 Variation diagrams of the major elements in terms of MgO (wt%) for different groups of rocks in Sierra Chica, adapted from Lagorio (2008) inciluding also ankaratrites. Diamonds: basanites of El Pungo (Bsn); filled circles: alkali basalt—trachyte suite (AkB-Tc); small circles: samples of the Sierra de los Cóndores and Los Molinos locality; *large circles* Almafuerte samples; *triangles* transitional basalt–latibasalt suite (TrB-Lb) in Sierra de Los Cóndores, Almafuerte and Despeñaderos localities; *square* ankaratrites from Los Molinos

Variation diagrams are consistent with fractionation of olivine, clinopyroxene and magnetite in both suites, as discussed in detail in Lagorio (2003, 2008). Within the alkaline suite, plagioclase and apatite removal is better supported from the late stages of the Almafuerte magmas, based on Sr and P_2O_5 lines of descent, respectively.

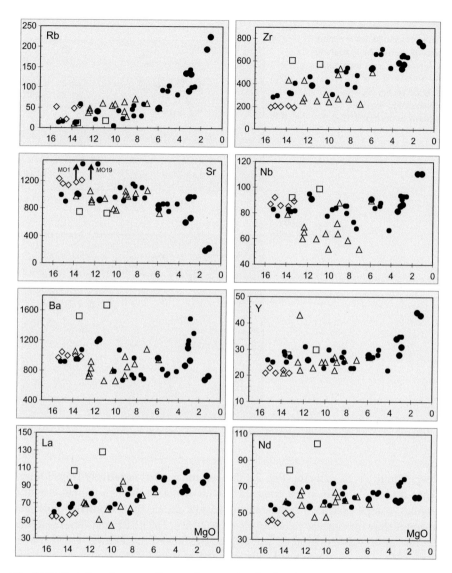

Fig. 2.14 Variation diagrams of trace elements (ppm) related to MgO (wt%). Adapted from Lagorio (2008) including also ankaratrites. Symbols as in Fig. 2.13

Relationship between incompatible elements (e.g. Zr vs. Th, Fig. 2.15a) both in alkaline and transitional suites shows no clear linear trends, suggesting that complete evolution must not be directly due only to simple crystal fractionation starting from primary magmas. The strong variation in the content of some incompatible elements (e.g. Zr) in the most primitive magmas is clearly visible in the

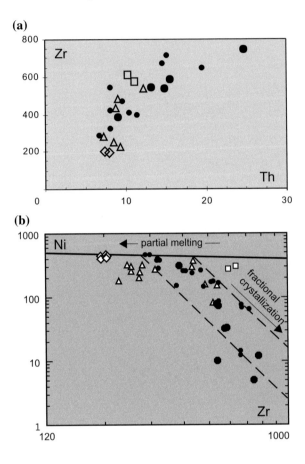

Fig. 2.15 Variation diagrams of trace elements for volcanic rocks of Sierra Chica, adapted by Lagorio (2008) including also ankaratrites. **a** Th versus Zr. **b** Zr versus Ni bilogarithmic diagram. Symbols as in Fig. 2.13

bilogarithmic Zr–Ni diagram (Fig. 2.15b), and the various negatively sloped linear trends that can be defined are compatible with evolution through various parental magmas, consistent with a heterogeneous mantle source. It is worthy to note that basanites show the lowest values of Zr (Fig. 2.15b), Hf, LREE and Sm, regarding the four distinguished groups, as well as the lowest La/Ta and La/Nb ratios (Fig. 2.16). On the other hand, alkaline and transitional basalts show variable contents of incompatible elements, particularly high for Zr (Fig. 2.15b), Hf and LREE in some transitional basalts; the latter are distinguished from the alkaline ones by their higher La/Ta, Th/Ta and La/Nb ratios (Fig. 2.16). Ankaratrites record the highest content of Zr (Fig. 2.15b), Ta, LREE, Sm, Hf, Ti and Tb, as well as high La/Ta ratio and the highest of La/Nb (Fig. 2.16).

In chondrite-normalized (cn; Boynton 1984) diagrams, the rocks of the alkaline suite have relationships $(La/Yb)_{cn}$ between 20 and 39 (Table 2.6; Lagorio 2008), showing patterns that diverge only for the trachyte of the Almafuerte locality (PM6; Fig. 2.17). The latter shows a steep LREE trend that tends to flatten out for HREE [$(Gd/Yb)_{cn}$ = 1.88] as well as a slight negative Eu anomaly (Eu/Eu* = 0.74),

Fig. 2.16 a La/Ta versus
Th/Ta and **b** Th/Ta versus
La/Nb for the most primitive
rocks of Sierra Chica (mg #
0.68–0.75) from the different
groups of lithologies of Sierra
Chica, adapted from Lagorio
(2008) also including
ankaratrites. *PM* primordial
mantle according to Sun and
McDonough (1989). Symbols
as in Fig. 2.13

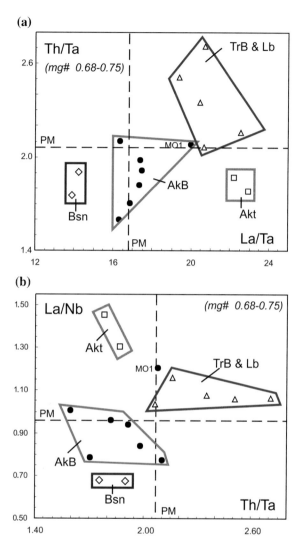

supporting fractionation of plagioclase. The transitional suite display (La/Yb)$_{cn}$ ratios between 25 and 41 (Table 2.6; Lagorio 2008), showing broadly similar patterns to those of the rocks of the alkaline suite, particularly rocks from Sierra de los Cóndores and Embalse Los Molinos (Fig. 2.17a). Basanites present a mean (La/Yb)$_{cn}$ of 28 (Table 2.6; Lagorio 2008) and also show patterns with no significant differences with respect to the latter ones (Fig. 2.17d). Ankaratrites display the highest (La/Yb)$_{cn}$ (average of 53, Table 2.6), with higher values for REE in comparison with the basanites, and the highest contents for LREE taken to account also the alkaline basalts (Fig. 2.17d).

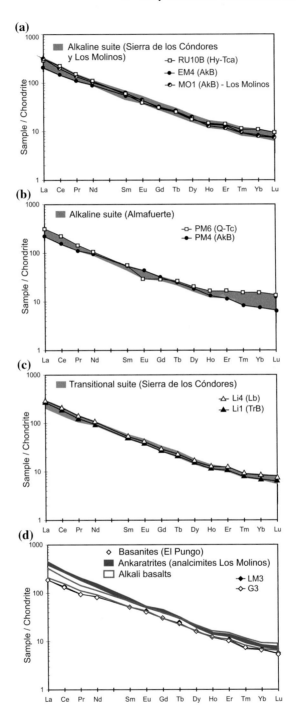

◀ **Fig. 2.17** Chondrite-normalized (Boynton 1984) REE diagrams for rocks of Sierra Chica, adapted from Lagorio (2008) also including ankaratrites. **a** Alkaline suite from Sierra de los Cóndores and Los Molinos locality. **b** Alkaline suite from Almafuerte. **c** Transitional suite from Sierra de los Cóndores. **d** Basanites from El Pungo and ankaratrites from Los Molinos, compared with alkali basalts. *Coloured fields* all samples analysed in each group

In multi-element diagrams normalized to the primordial mantle (Sun and McDonough 1989), rocks of the alkaline suite show negative anomalies in Rb and K for the less evolved samples, showing a gradual increase in the contents of Rb, Th, K, Zr and Hf towards the Hy- and Q-normative trachyandesites and trachytes (Fig. 2.18a, b). Q-varieties are also characterized by negative Sr, P and Ti anomalies, more pronounced in trachytes (Fig. 2.18b). It should be noted that alkaline basalts of dykes from Los Molinos are distinguished by presenting a positive anomaly for Sr, in contrast to the clearly negative anomaly that characterizes all samples from Sierra de los Cóndores. Transitional suite shows similar patterns regarding the less evolved samples of the alkaline suite from Sierra de los Cóndores (Fig. 2.18c). However, a slightly negative spike of Ta is shown, particularly conferred by the most evolved latibasalt (Fig. 2.18c). Basanites patterns are similar to those of the most primitive rocks of both suites, though they clearly present the lowest contents of Zr and Hf as well as low tenors of REE (Fig. 2.18d). Ankaratrites record the highest values of Ta, LREE, P, Sm, Zr, Hf, Ti and Tb; they also display a clear negative anomaly for Sr, in contrast to what happens in relation to the other rocks (alkaline basalts) of the same locality (Los Molinos) (Fig. 2.18d).

2.1.1.7 Petrogenesis

Differentiation Processes

In order to test fractional crystallization quantitatively, mass balance calculations based on major elements (Stormer and Nicholls 1978) were carried out. Rayleigh's trace element contents were calculated using mass balance results and the partition coefficients employed by Lagorio (2003, 2008) and shown in Tables 2.7 and 2.8.

Some transitions appear compatible (sum of the squares of major element residua, $\Sigma res^2 < 1.0$) with simple fractional crystallization (Table 2.8), e.g. from EM3 (trachybasalt) to RU7 (trachyandesite), from PM1 (trachyandesite) to PM7 (trachyphonolite) and from Li1 (transitional basalt) to Li4 (latibasalt). The calculated/observed trace element ratios also are in general satisfactory (0.90–1.35; Table 2.8), except for Cr and Ni (0.28–2.10), and for Rb and Ba (0.57–1.77). Ni and Cr discrepancies may be partly due to uncertainties on partition coefficients, while those for Rb and Ba may probably reflect alteration. On the contrary, other transitions, e.g. from EM4 (alkali basalt) to EM3 (trachybasalt), similarly compatible with fractional crystallization in terms of major element results (though a higher Σres^2 is obtained, Table 2.8), reveal many trace element ratios quite different from

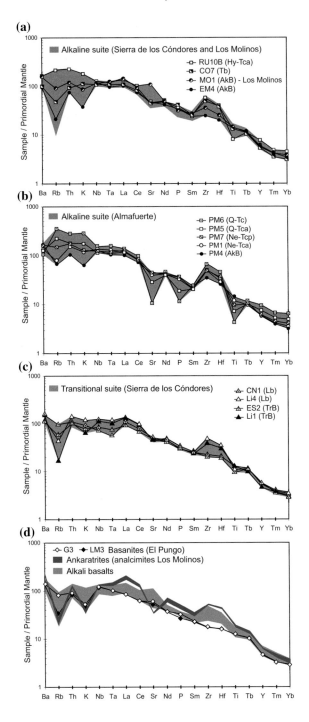

◄ **Fig. 2.18** Primordial mantle (Sun and McDonough 1989) normalized multi-elemental plots for rocks of the Sierra Chica, adapted from Lagorio (2008) also including ankaratrites. **a** Alkaline suite in Sierra de los Cóndores and Los Molinos. **b** Alkaline suite from Almafuerte. **c** Transitional suite in Sierra de los Cóndores. **d** Basanites of El Pungo and ankaratrites of Los Molinos dam, compared with alkali basalts. *Coloured fields* all samples analysed in each group

1.0 (up to 1.96). It should be also noted that the transitional basalt to latibasalt transition is only supported starting from basalts with low normative hypersthene (<2 %, e.g. Li1, CN14).

Quantitative evidence therefore also indicates that simple fractional crystallization does not completely account for the SCC magmatic differentiation, suggesting therefore distinct parental magmas and/or evolution in an open-system magma chamber.

Actually, this is congruent with trace element variation diagrams. The remarkable variation in some IE contents (e.g. Zr) of the most primitive rocks is clearly seen in a Zr–Ni log–log diagram (Fig. 2.15b), and the diverse linear trends with negative slope (that can be defined) are more compatible with an evolution through different parental magmas.

Although there is some petrographic evidence of upper crustal contamination in some rocks (particularly in dykes from Los Molinos and in a few lavas of Sierra de los Cóndores, as previously stated by Gordillo and Lencinas 1967a, 1969), the lack of a significant Ta–Nb negative depletion in the multi-elemental diagrams may indicate that crustal contamination was not a relevant a process (Fig. 2.18, only latibasalt CN1 displays a slight anomaly). In a $(La/Nb)_{PM}$ versus K_2O plot (Fig. 2.19a), this sample also appears close to the lower crust ratio of Rudnick and Gao (2003); considering

Table 2.7 Mineral/liquid partition coefficients used for fractional crystallization modelling

	Ol		Cpx		Pl		Mt		Ap	
	P	E	P	E	P	E	P	E	P	E
Cr	2.78	2.78	5.69	5.69	0.03	0.03	4.21	7.70	0.01	0.01
Ni	15.5	15.5	2	2	0.03	0.03	2.39	6.54	0.01	0.01
Ba	0.09	0.09	0.07	0.01	0.63	3.90	0.14	0.14	0.95	0.95
Rb	0.08	0.08	0.1	0.01	0.03	0.20	0.08	0.08	0.01	0.01
Sr	0.01	0.01	0.33	0.1	2.5	4.41	0.16	0.16	1.67	1.67
La	0.01	0.01	0.22	0.1	0.12	0.46	0.53	0.53	5.16	5.16
Ce	0.01	0.01	0.34	0.20	0.14	0.36	0.56	0.56	6.34	6.34
Nd	0.01	0.01	0.68	0.35	0.07	0.31	0.55	0.55	6.6	6.6
Zr	0.05	0.01	0.24	0.24	0.01	0.01	0.35	0.35	0.01	0.01
Y	0.01	0.01	0.77	0.77	0.05	0.24	0.55	0.55	5.08	5.08
Nb	0.17	0.01	0.12	0.12	0.01	0.01	3.86	3.86	0.01	0.01
Th	0.04	0.04	0.13	0.13	0.02	0.16	0.18	0.11	0.95	0.95

Ol olivine, *Cpx* clinopyroxene, *Pl* plagioclase, *Mt* magnetite, *Ap* apatite. Data from Caroff et al. (1993), De Min (1993) and Comin-Chiaramonti et al. (1997). *P* and *E* primitive and evolved magmas, respectively

Table 2.8 Representative mass balance calculations for major elements (Stormer and Nicholls 1978) and trace elements (Rayleigh equation) for fractional crystallization within AkB-Tc and TrB-Lb suites, according to Lagorio (2008)

	EM4 AkB	EM3 Tcb	Ol(90)	Cpx	Mt	Ap	EM3 calc.
SiO_2	46.82	51.08	41.02	48.53	0.04	0.00	51.15
TiO_2	2.90	2.32	0.00	2.6	22.67	0.00	2.50
Al_2O_3	9.71	14.21	0.02	4.86	2.57	0.00	14.38
FeO_t	10.61	8.78	9.45	6.11	72.88	0.00	8.79
MnO	0.16	0.14	0.15	0.11	1.12	0.07	0.13
MgO	15.30	7.61	49.31	13.71	0.61	0.00	7.65
CaO	9.89	8.05	0.05	23.61	0.1	57.01	8.03
Na_2O	2.73	4.06	0.00	0.48	0.00	0.00	4.27
K_2O	1.13	3.24	0.00	0.00	0.00	0.00	2.44
P_2O_5	0.75	0.52	0.00	0.00	0.00	42.92	0.66
Mineral (wt%)			40.06	49.29	8.92	1.73	
F (%)	59.64			$\sum res^2 = 0.78$			

	EM4	EM3	EM3 FC	Calc./obs.	EM3 AFC	Calc./obs.
Cr	653.00	282.00	119.00	0.42	95.00	0.34
Ni	464.00	152.00	17.00	0.11	12.00	0.08
Rb	14.00	30.00	22.00	0.73	29.00	0.97
Ba	918.00	745.00	1462.00	1.96	1478.00	1.98
Sr	994.00	1100.00	1495.00	1.36	1495.00	1.36
Nb	83.00	73.00	109.00	1.49	107.00	1.46
Zr	287.00	376.00	441.00	1.17	445.00	1.18
Y	26.00	23.00	33.00	1.43	34.00	1.46
La	65.30	71.90	96.30	1.34	96.40	1.34
Ce	121.00	135.10	171.00	1.27	170.90	1.26
Nd	56.10	53.90	72.60	1.35	71.80	1.33
Th	6.60	9.03	10.44	1.16	10.52	1.17

	EM3 Tcb	RU7 Tca	Ol(86)	Cpx	Pl(46)	Mt	Ap	RU7 Calc.
SiO_2	51.08	56.07	40.19	51.78	56.86	0.27	0.00	55.84
TiO_2	2.32	2.15	0.00	1.16	0.2	21.14	0.00	2.09
Al_2O_3	14.21	14.97	0.00	2.36	26.88	1.69	0.00	14.84
FeO_t	8.78	6.74	12.88	5.4	0.46	75.28	0.00	6.60
MnO	0.14	0.13	0.17	0.09	0.00	1.03	0.07	0.13
MgO	7.61	5.13	46.51	15.94	0.00	0.17	0.00	5.11
CaO	8.05	5.31	0.25	22.88	8.91	0.42	57.01	5.35

	EM3	RU7	RU7 FC	Calc./obs.
Cr	282.00	88.00	136.00	1.55
Ni	152.00	70.00	45.00	0.64
Rb	30.00	92.00	52.00	0.57
Ba	745.00	738.00	1079.00	1.46
Sr	1100.00	863.00	903.00	1.05
Nb	73.00	86.00	99.00	1.15
Zr	376.00	670.00	637.00	0.95

(continued)

Table 2.8 (continued)

	EM3	RU7	Ol(86)	Cpx	Pl(46)	Mt	Ap	RU7		EM3	RU7	RU7	Calc./obs.
	Tcb	Tca						Calc.				FC	
Na₂O	4.06	3.77	0.00	0.39	6.15	0.00	0.00	4.39	Y	23.00	28.00	33.00	1.18
K₂O	3.24	5.11	0.00	0.00	0.54	0.00	0.00	5.40	La	71.90	84.00	107.90	1.28
P₂O₅	0.52	0.61	0.00	0.00	0.00	0.00	42.92	0.26	Ce	135.10	161.00	194.00	1.20
Mineral (wt%)			14.92	23.73	47.64	10.91	2.79		Nd	53.90	60.30	75.00	1.24
F (%)	55.41		∑res² = 0.70						Th	9.03	14.50	15.43	1.06

	PM1	PM7	Ol(82)	Cpx	Pl(51)	Mt	Ap	PM7		PM1	PM7	PM7	Calc./obs.
	Tca	Tcp						calc.				FC	
SiO₂	52.36	55.29	39.52	50.36	56.16	0.11	0.00	55.25	Cr	81.00	20.00	42.00	2.10
TiO₂	2.59	2.05	0.04	1.73	0.14	29.87	0.00	2.14	Ni	73.00	10.00	25.00	2.50
Al₂O₃	15.47	17.59	0.05	3.54	27.21	1.35	0.00	17.66	Rb	51.00	92.00	63.00	0.68
FeOₜ	8.30	6.49	16.50	8.57	0.46	66.38	0.00	6.42	Ba	972.00	1109.00	1029.00	0.93
MnO	0.11	0.10	0.21	0.18	0.00	2.17	0.07	0.05	Sr	846.00	949.00	852.00	0.90
MgO	5.95	3.12	43.39	13.76	0.00	0.03	0.00	3.13	Nb	91.00	86.00	102.00	1.19
CaO	6.20	4.76	0.23	21.18	9.36	0.09	57.01	4.76	Zr	543.00	539.00	660.00	1.22
Na₂O	4.70	6.13	0.04	0.68	5.33	0.00	0.00	5.68	Y	27.00	28.00	30.00	1.07
K₂O	3.59	3.85	0.02	0.00	1.35	0.00	0.00	4.32	La	80.60	89.40	93.30	1.04
P₂O₅	0.72	0.62	0.00	0.00	0.00	0.00	42.92	0.58	Ce	144.70	158.00	165.30	1.05
Mineral (wt%)			26.93	38.7	19.39	11.89	3.09		Nd	56.30	61.30	63.50	1.04
F (%)	79.70		∑res² = 0.44						Th	13.20	14.80	16.06	1.09

	Li1	Li4	Ol(90)	Cpx	Mt	Ap	Li4		Li1	Li4	Li4	Calc./obs.
	TrB	Lb					Calc.				FC	
SiO₂	48.55	52.70	41.02	50.70	0.28	0.00	52.50	Cr	492.00	270.00	125.00	0.46

(continued)

Table 2.8 (continued)

	Li1	Li4	Ol(90)	Cpx	Mt	Ap	Li4		Li1	Li4	Li4	Calc./obs.
	TrB	Lb					Calc.				FC	
TiO_2	2.77	2.31	0.00	0.64	30.52	0.00	2.35	Ni	366.00	151.00	42.00	0.28
Al_2O_3	10.37	13.45	0.02	4.53	0.73	0.00	13.97	Rb	11.00	28.00	16.00	0.57
FeO_t	9.85	7.91	9.45	7.32	66.04	0.00	7.84	Ba	1057.00	852.00	1509.00	1.77
MnO	0.15	0.14	0.15	0.27	0.69	0.07	0.11	Sr	974.00	978.00	1316.00	1.35
MgO	13.87	8.98	49.31	13.40	1.38	0.00	9.10	Nb	79.00	88.00	95.00	1.08
CaO	9.60	7.25	0.05	22.58	0.36	57.01	7.41	Zr	436.00	538.00	605.00	1.12
Na_2O	2.22	2.91	0.00	0.55	0.00	0.00	3.08	Y	21.00	25.00	24.00	0.96
K_2O	1.96	3.61	0.00	0.00	0.00	0.00	3.15	La	91.60	93.40	119.80	1.28
P_2O_5	0.65	0.75	0.00	0.00	0.00	42.92	0.48	Ce	165.00	174.00	207.00	1.19
Mineral (wt%)			32.64	53.83	10.57	2.95		Nd	62.60	66.00	72.80	1.10
F (%)	67.03					$\sum res^2 = 0.67$		Th	8.76	12.10	12.41	1.03

AFC modelling (De Paolo 1981) for AkB-Tcb transition, with $r = 0.1$ and mean composition of Córdoba basement (Rapela et al. 1998) as the contaminant *It can be stop instead of full stop, so this paragraph could be immidiately after the other.AkB* alkali basalt, *Tcb* trachybasalt, *Tca* trachyandesite, *Tcp* trachyphonolite. Mineral phases, with their respective compositions, subtracted from the parent liquid: *Ol* olivine, *Cpx* clinopyroxene, *Mt* magnetite, *Pl* plagioclase, *Ap* apatite. In parentheses: forsterite and anorthite contents of olivine and plagioclase, respectively; *calc.* calculated composition, *obs.* observed composition. *Mineral (wt%)* amount of the subtracted minerals, *F (%)* weight fraction of residual liquid. $\sum res^2$ sum of squares of major element residuals. *FC* fractional crystallization; *AFC* assimilation and fractional crystallization. EM4 is similar to EM5 and comes from the same level (picritic basalts of Gordillo and Lencinas), so that mineral phases employed in the modelling are those measured in EM5

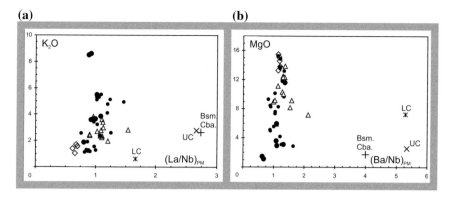

Fig. 2.19 **a** (La/Nb)$_{PM}$ ratios versus K$_2$O (wt%). **b** (Ba/Nb)$_{PM}$ ratio versus MgO (wt%) for the volcanic rocks of SCC, based on Lagorio (2008). *PM* primordial mantle normalization (Sun and McDonough 1989); Bsm. *Cba* Córdoba basement (Rapela et al. 1988); *UC* and *LC* upper and lower crust, respectively, from Rudnick and Gao (2003)

other ratios such as (Ba/Nb)$_{PM}$ (all samples display ratios < 2; Fig. 2.19b), and La/Ta (mean value = 19), crustal interaction does not appear as remarkable.

It should also be noted that no significant discrepancies in calculated/observed Th contents (typical upper crustal immobile element) appear in the Rayleigh crystal fractionation model (Table 2.8), despite some controversy in the partition coefficients. Anyway, AFC modelling (assimilation and fractional crystallization; De Paolo 1981) was also performed for the EM4–EM3 transition, considering the mean composition of the Córdoba crystalline basement (Rapela et al. 1998) as the contaminant. Calculations show in this case that this model does not give a satisfactory quantitative solution either (Table 2.8).

From six Sr–Nd isotopic data from Sierra de los Cóndores provided by Lucassen et al. (2002), five have ƐSr (125 Ma) from 15.1 to 24.1 and ƐNd (125 Ma) between −3.5 and −6.1; only one sample indicates significant crustal contamination (ƐSr = 50.8 and ƐNd = −5.1). Nevertheless, Pb isotopes also obtained by the latter authors point out <10 % contamination with lower crust, not detectable through Sr–Nd isotopes.

Textural evidence of solid–liquid disequilibrium from some plagioclase and clinopyroxene macro- and phenocrysts, consistent with frequent slight core to rim Ca and Mg increments, and particularly with the common Mg increase from clinopyroxene micro- and phenocryst rims towards groundmass cpx-microlites, suggests interaction with less evolved magma batches (e.g. Shimizu and Le Roex 1986; O'Brien et al. 1988), as in the RTF model of O'Hara and Mathews (1981). Therefore, the ocellar textures of some alkali basalts and trachybasalts reflecting local liquid immiscibility (e.g. Philpotts 1976; Shelley 1993) are also in accordance with mechanical interaction between magmas. This type of process is also reflected by the pegmatoid alkaline gabbroid segregations that occur in some basanitic rocks of southern Sierra de los Cóndores, as described by Escayola et al. (1998), bearing analogies with those reported in younger volcanic rocks of the Chaján locality by Galliski et al. (1996, 2004).

In summary, the available data suggest that the SCC magma evolution must have taken place at crustal level(s) from distinct parental melts, mainly through fractional crystallization in an open-system magma chamber, particularly involving local magma mixing with more primitive batches, besides some amount of crustal contamination.

Mantle Source

Some of the more magnesium-rich rocks of SCC can be considered primary magmas due to their high mg# values (>0.68) and Ni contents (>200 ppm), as it is shown in Table 2.6. Representative samples are the basanite G3 (mg# 0.74, Ni = 443 ppm), the alkali basalts such as EM4, MO1 and PM4 (mg# 0.75, 0.74 and 0.69; Ni = 464, 249 and 311 ppm, respectively) and the transitional basalts such as Li1 and CN14 (mg# 0.74 and 0.73, Ni = 366 and 314 ppm, respectively), as described by Lagorio (2003, 2008). The latter reveal primary magmas, in equilibrium with olivine from the mantle (Fo_{89-90}), also consistent with the presence of dunitic and/or spinel lherzolite xenoliths. $(La/Yb)_{cn}$ values are high (24–41), strongly supporting garnet in the peridotite residua, as stated by Kay and Ramos (1996) for magmas from Los Molinos and Almafuerte localities. These latter authors pointed out that some IE ratios (e.g. La/Ta, La/Ba) of rocks from those localities were characteristic of OIB magmas, also supported by the isotopic data. Therefore, they characterized a garnet-bearing OIB-like mantle source for magmas from Sierra Chica de Córdoba. Such a mantle source was after also outlined for magmas of Sierra de los Cóndores by Sánchez and Bermúdez (1997). Garnet in the mantle source was confirmed by the presence of garnet-bearing xenoliths in basanites from southern Sierra de los Cóndores pointed out by Escayola et al. (1998).

Lagorio (2003, 2008), considering geochemical data from the entire Sierra Chica, pointed out that low Al_2O_3 content (<15 %) together with high olivine (20–26 %) and variable nepheline (3–15 %) or low orthopyroxene (≤1) CIPW values suggest high depths of melting (Green 1970), notionally towards the garnet–spinel transition (about 26 kbar, near 100 km depth; Takahashi and Kushiro 1983).

On the other hand, thermobarometric studies in the garnet-bearing lherzolite xenoliths hosted in basanitic lavas from southern Sierra de los Cóndores allowed Escayola et al. (1998, 1999) to estimate depths of up to 160 km in the mantle for the generation of such magmas.

Furthermore, negative anomalies for K and Rb in multi-element diagrams (Fig. 2.18) are consistent with the presence of residual phlogopite in the source.

The melting degrees for the primary magma genesis were calculated by mass balance (Stormer and Nicholls 1978) using the major element compositions of an enriched peridotite. Calculations were carried out for both dry and hydrous (phlogopite) garnet peridotites (P1 and P2, respectively, Table 2.9).

Results indicate that similar degrees of melting are necessary to match the magma compositions of the basanite, the alkali and transitional basalts. Melting

Table 2.9 Anhydrous and hydrous peridotite mineral assemblages calculated (Stormer and Nicholls 1978) from chemical compositions of Ringwood (1966; P1 and P2) and calculated mineral assemblages of mantle residua after extraction of G3, EM4, MO1, PM4, Li1, and CN14 primary magmas

	P1	G3	EM4	MO1	PM4	Li1	CN14
	Initial	Resid.	Resid.	Resid.	Resid.	Resid.	Resid.
OL[a] (wt%)	54.88	58.40	59.52	60.51	59.54	61.25	62.04
OPX[b]	18.83	20.64	19.78	19.14	19.03	18.82	17.59
CPX[c]	11.14	8.64	8.66	8.63	9.38	8.51	9.25
GT[b]	15.14	12.32	12.05	11.72	12.05	11.42	11.12
Σres^2	0.53	0.41	0.38	0.35	0.38	0.36	0.34
F (%)		6.38	7.19	7.39	5.84	8.20	7.91
	P2	G3	EM4	MO1	PM4	Li1	CN14
	Initial	Resid.	Resid.	Resid.	Resid.	Resid.	Resid.
OL[a] (wt%)	54.49	57.51	58.66	59.00	58.48	59.63	60.26
OPX[b]	18.12	19.71	19.00	18.59	18.37	18.28	17.36
CPX[c]	11.37	9.23	9.13	9.34	9.84	9.23	9.80
GT[b]	13.16	10.92	10.50	11.13	10.80	10.68	10.60
PHL[d]	2.86	2.64	2.70	1.94	2.50	2.19	1.97
Σres^b	0.40	0.31	0.28	0.30	0.30	0.30	0.29
F (%)		5.48	6.47	5.84	4.99	6.61	6.35

Mantle mineral compositions: *OL* olivine, *OPX* orthopyroxene, *CPX* clinopyroxene, *GT* garnet, *PHL* phlogopite, according to: [a]De Min (1993), [b]MacGregor (1974), [c]Bristow (1984), [d]Wilkinson and Le Maitre (1987). Σres^2 sum of squares of major element residuals; *F* melting degree; *resid.* residua

degrees are between 6–8 % and 5–7 % for dry and hydrous sources, respectively (Table 2.9). It is clear that if the modelling is performed from a peridotite less enriched in incompatible elements, percentages of melting do not exceed 2 %. In fact, it should be noted that, previously, Kay and Ramos (1996) obtained degrees of melting <2 % from the magmas represented by the analcimites of Los Molinos.

Incompatible trace elements (IE) in the peridotitic sources were calculated, assuming batch melting (Hanson 1978) from pyrolite composition, considering the partition coefficients used by Comin-Chiaramonti et al. (1997) and the calculated melting degrees and solid residua (Table 2.9). As K- and Rb-negative spikes of the primary magmas IE- patterns support phlogopite as a residual phase in the mantle (Fig. 2.20), a hydrous source was therefore preferred.

The IE contents in the initial sources were calculated from the primary basanite, alkali and transitional basalts and plotted in multi-elemental diagrams (Fig. 2.20). Their enrichment relative to the primitive mantle of Sun and McDonough (1989) may be up to 10–20 times (e.g. for Ba and K).

In summary, our calculations indicate that the slight compositional features and differences concerning basanite, alkaline and transitional suite primary melts mainly reflect heterogeneous lithospheric mantle sources.

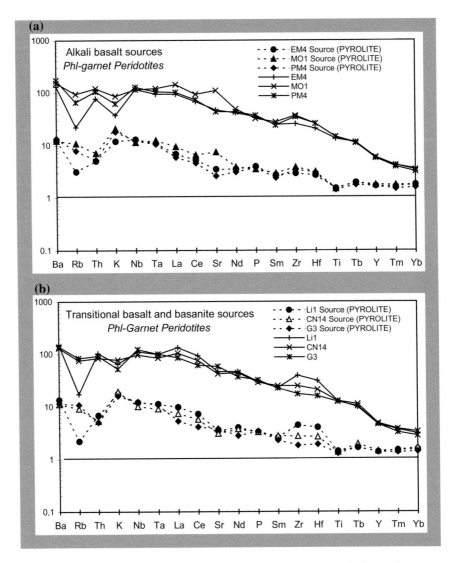

Fig. 2.20 Multi-elemental plots of calculated (batch melting; Hanson 1978) mantle sources, normalized to primordial mantle (PM) of Sun and McDonough (1989) for distinct primary magmas of Sierra Chica de Córdoba, starting from an enriched peridotite (pyrolite; Ringwood 1966), as shown in Lagorio (2008). **a** Garnet phlogopite mantle sources of primary alkali basalts and **b** garnet phlogopite mantle sources of primary transitional basalts and basanite. Each figure also includes multi-element plots for the respective primary magmas

2.1.1.8 Comparison with Alkaline Volcanism Around Paraná Basin

Distinction in high- and low-Ti types of Paraná (Bellieni et al. 1984) and Gondwana (Cox 1988) basalts was also adopted by Gibson et al. (1996) to discriminate the

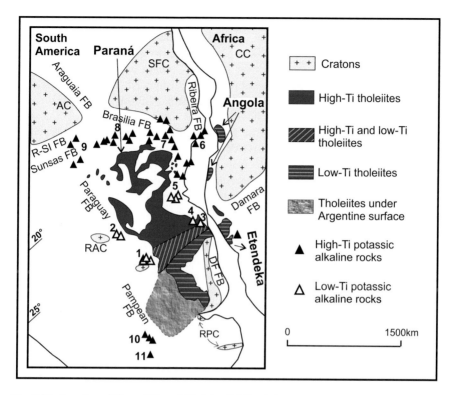

Fig. 2.21 Location of Sierra Chica of Córdoba (SCC) relative to other high-Ti and low-Ti mafic potassic rocks and continental flood basalts across western and central Gondwana, modified from Piccirillo and Melfi (1988), Gibson et al. (1996), Marzoli et al. (1999) and Lagorio (2008). *AC* Amazonian craton, *SFC* San Francisco craton, *CC* Congo craton, *RAC* Río Apa craton, *RPC* Río de La Plata craton; *FB* fold belt, *DF FB* Dom Feliciano fold belt, *R-SI FB* Rondonia-San Ignacio fold belt, *1* eastern Paraguay, *2* Amambay, *3* Anitápolis, *4* Lages, *5* Ponta Grossa, *6* Serra do Mar, *7* Alto Paranaíba Igneous Province, *8* Goiás Alkaline Province, *9* Poxoréu, *10* Sierra Chica of Córdoba, *11* Chaján

mafic potassic rocks related to Paraná basin (Figs. 2.21). Low-Ti potassic alkaline rocks outcrop along the western (e.g. eastern Paraguay, Amambay) and eastern (e.g. Lages, Anitápolis, Ponta Grossa) borders of the central sector of the Paraná basin (Fig. 2.21). Instead, the high-Ti types are located in the northern and eastern margins of the northern part of the basin (e.g. Alto Paranaíba, Goiás, Serra do Mar; Fig. 2.21). The potassic rocks of Sierra Chica of Córdoba (SCC) are of high Ti and are located in the south-western edge of the Paraná basin (in fact, the Chaco-Paraná basin; Figs. 1.1a, b and 2.21), as pointed out by Lagorio (2008). Similarly, Late Cretaceous rocks from the locality of Chaján (southern Córdoba) are also potassic as well as high Ti and have a peripheral location with respect to the tholeiitic volcanic flows of Paraná, though they are slightly more distant, about 150 km south-west of the Sierra Chica (Figs. 1.1a and 2.21).

As it is well-known from the literature, low- and high-Ti potassic rocks around the Paraná basin greatly differ in terms of K, Nb, Ta and La relationships, as pointed out by several authors. The low-Ti rocks have IE patterns with negative spikes (e.g. eastern Paraguay; La/Nb$_{PM}$ higher than 1.6, e.g. Comin-Chiaramonti et al. 1997, 2013), whereas the high-Ti ones do not reveal any distinct Nb-Ta anomaly, or may even show a slight positive spike (e.g. Alto Paranaíba; Gibson et al. 1995).

Comparison of the most primitive magmas of Sierra Chica of Córdoba (SCC) shows similar patterns to those from the Alto Paranaíba igneous province (APIP) and Goias alkaline province (GAP), as well as the ones from the locality of Chaján (Fig. 2.22a). However, rocks from APIP have higher contents of LIL and LRE elements, and also La/Yb ratios (normalized relations up to 155), according to the ultrapotassic nature that characterizes that province (Gibson et al. 1995, Carlson et al. 1996, 2007). By contrast, low-Ti rocks (e.g. eastern Paraguay, Anitápolis) have patterns distinctively characterized by a negative anomaly for Nb–Ta (Fig. 2.22b). According to Comin-Chiaramonti et al. (1997, 2013), these compositional differences might reflect different IE enrichments, therefore metasomatic events, that have occurred in Early–Middle Proterozoic (low Ti: 2.0 to 1.5 Ga) and Late Proterozoic (high-Ti: 1.0 to 0.5 Ga), respectively, from Nd model ages presented by diverse authors. Gibson et al. (1996) also pointed out that the low-Ti magmas are associated with the cratonic regions, while the high-Ti ones are located in Proterozoic mobile belts.

On the other hand, the initial isotopic ratios of Sr and Nd samples from the Sierra Chica of Córdoba presented by Kay and Ramos (1996) and Lucassen et al. (2002) are plotted next to the composition of enriched mantle I (EMI), as well as the high-Ti alkaline localities APIP and GAP (Fig. 2.23). The same is observed for the Chaján rocks, as shown in the figure, from data presented by Lucassen et al. (2002). These authors, on the basis of Pb isotopes characterized an enriched mantle, similar to the ancient mantle located beneath the Brazilian shield, but with distinguishing characteristics between Sierra de los Cóndores (southern Sierra Chica) and Chaján, as the high $^{208}Pb/^{204}Pb$ ratios of rocks from the former are clearly more similar to those of the alkaline provinces of Paraná. Lucassen et al. (2002) also pointed out the significant isotopic difference between the source of the magmas of Córdoba and that of the Salta Group (northern Argentina; Fig. 2.22a). The latter volcanism is mainly sodic and of Late Cretaceous age; its source is clearly depleted, as it has been previously reported by Kay and Ramos (1996) from Sr and Nd isotopic data. Morover, Lucassen et al. (2007) reinforced this evidence, also for most of the Cretaceous alkaline mafic intra-plate rocks along the back arc of the Central Andes, as they were derived from a depleted lithospheric mantle according to a complete isotopic study.

The new $^{40}Ar/^{39}Ar$ dating of SCC of 129.6 ± 1.0 Ma is in agreement with geochronological data obtained so far (see Table 2.1), indicating that this volcanism is Early Cretaceous in age and slightly posthumous to the large tholeiitic event of PMP.

In relation to the age of other potassic localities around the Paraná basin, as central–eastern Paraguay and Anitápolis (Brazil), a comparison must be performed

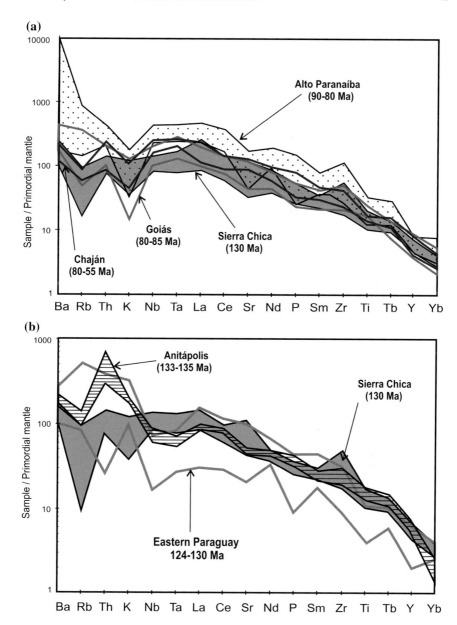

Fig. 2.22 Primordial mantle (Sun and McDonough 1989) normalized multi-element plots for the most primitive rocks (mg# 0.68–0.75) of Sierra Chica of Córdoba in comparison with: **a** high-Ti mafic potassic rocks of Alto Paranaíba and Goiás Provinces from Brazil (Gibson et al. 1995, 2006; Carlson et al. 1996, 2007) and Chaján locality (Córdoba, Argentina; Viramonte et al. 1994; Galliski et al. 1996; Quenardelle and Montenegro 1998) and **b** low-Ti less evolved potassic rocks of eastern Paraguay (Comin-Chiaramonti et al. 1997, 2007, 2013) and Anitápolis (Brazil; Comin-Chiaramonti et al. 1999). Modified from Lagorio (2008) and Lagorio et al. (2014)

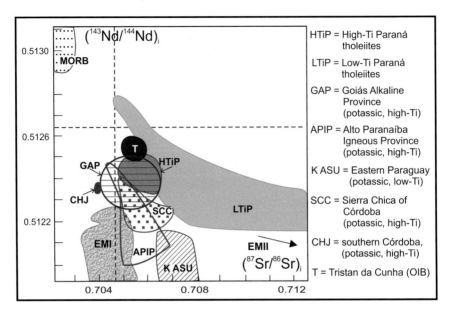

Fig. 2.23 Initial $^{87}Sr/^{86}Sr$ versus $^{143}Nd/^{144}Nd$ of the alkaline volcanic rocks of Sierra Chica of Córdoba (SCC; data: Kay and Ramos 1996; Lucassen et al. 2002) compared to other alkaline potassic localities, as Chaján (CHJ; data: Lucassen et al. 2002) and others peripheral to the Paraná Magmatic Province (PMP), as the Alto Paranaíba Igneous Province (APIP; Gibson et al. 1995; Carlson et al. 1996, 2007), Goiás Alkaline Province (GAP; Carlson et al. 1996, 2007) and eastern Paraguay (K ASU; Comin-Chiaramonti et al. 1997, 2007), together with those of high-Ti and low-Ti tholeiites from Paraná Magmatic Province (PMP; data: Piccirillo and Melfi 1988; Peate and Hawkesworth 1996; Peate 1997), Tristan da Cunha (TC; data: le Roex et al. 1990), MORB, EMI and EMII of Zindler and Hart (1986). Based on Lagorio et al. (2014)

taking into account the latest data given by diverse authors. For eastern Paraguay, weighted mean, $^{40}Ar/^{39}Ar$ dating on phlogopite is 127.56 ± 0.45 Ma, from samples that gave the best spectra according to Gibson et al. (2006); using the new age standard of Kuiper et al. (2008) the date is 128.39 Ma. Otherwise, $^{40}Ar/^{39}Ar$ dating performed by other authors span from 129.0 ± 2 Ma to 123.6 ± 0.5 Ma (Comin-Chiaramonti et al. 2007), and after applying the above-mentioned standard, they range between 129.8 and 124.4 Ma. On the other hand, Anitápolis yielded an age between 135 and 132 Ma, according to Gibson et al. (2006) with Kuiper's recalculation of (2008). Therefore, SCC volcanic event is partially coeval with respect to eastern Paraguay one and younger than that of Anitápolis, beyond that all of them are Early Cretaceous in age.

By contrast, SCC volcanism is older than high-Ti potassic localities in Brazil (e.g. Alto Paranaíba, Goiás and Serra do Mar), as the latter are of Late Cretaceous/Palaeogene age according to diverse authors (e.g. Gibson et al. 1995, Carlson et al. 2007). Likewise, rocks from Chaján and Estancia Guasta (Figs. 1.1a and 2.21, Córdoba Province, Argentina) were formed contemporaneously with those of the Brazilian localities, taken into account ages provided by other authors (e.g. Valencio et al. 1980; Gordillo et al. 1983), and all of them are clearly posthumous with respect to the Paraná Magmatic Province.

SCC volcanism, as other alkaline localities from SE Brazil, is located over a mobile belt; in this case, the Pampean belt, coeval with Brazilian orogens (e.g. Paraguay and Araguaia fold belts), considered as Late Proterozoic–Early Cambrian (e.g. Escayola et al. 2007) or Early–Mid-Cambrian in age (e.g. Rapela et al. 2007, 2011, Thover et al. 2010; Casquet et al. 2012), whereas the APIP and GAP volcanism are placed on the Brasilia mobile belt. It should be noted that subduction-related granitoids outcrop north of the Sierra Chica (in Sierra Norte of Córdoba; Lira et al. 2014), with ages of 537 ± 4 Ma and 530 ± 4 Ma (Iannizzotto et al. 2013); they are associated with Cambrian to Ordovician volcanic and sub-volcanic bodies of calc-alkaline nature (O'Leary et al. 2014). Nevertheless, the volcanic rocks of SCC do not reveal chemical parameters that can be easily related to slab-derived mantle metasomatism, as the lack of a negative Nb-Ta anomaly, low La/Nb (0.6–1.2) and Ba/Nb (9–13) ratios, La/Ta < 25 and Ba/La < 20, as in rocks from high-Ti localities of Brazil. Therefore, this ancient subduction must have not contaminated the source of the magmas of SCC; precisely Nd model ages obtained by Lucassen et al. (2002) vary between 1.22 and 0.96 Ga, suggesting a reactivation of lithospheric mantle modified during Mesoproterozoic times. While the age of metasomatism in the lithospheric mantle of APIP and GAP is not exactly defined, isotopic data allowed to infer that it must have taken place during the configuration of the Brasilia mobile belt in the Middle to Late Proterozoic (Carlson et al. 2007). Therefore, the metasomatic events affecting the mantle sources of SCC, GAP and APIP seem to have been at least partially coeval.

Geochemical features reflect small-scale heterogeneity of SCC lithospheric mantle source (Lagorio 2008). This might be more related to metasomatic events involving volatile-rich small-volume melts from the asthenosphere, as indicated by the occurrence of trapped primary CO_2 inclusions and bleb-like glass in olivines from spinel peridotite nodules sampled by SCC alkali basalts (Lagorio and Montenegro 2004). Also, the study of garnet-bearing mantle xenoliths from southern Sierra de los Cóndores let Escayola et al. (1998, 1999) describe metasomatic processes in the lithospheric mantle, which is in agreement with most of the reports from the peripheral localities around Paraná basin (e.g. Gibson et al. 1995, 2006; Comin-Chiaramonti et al. 1997, 2007, 2013).

2.1.2 Early Cretaceous in the Levalle Basin of Córdoba Province

In southern Córdoba Province, there are other Lower Cretaceous rift deposits that belong to the Levalle basin (Fig. 2.2a), though they are presently buried. This basin is probably composed of three megasequences according to geophysical information and a well that was drilled to test its petroleum potential (Webster et al. 2004; Chebli et al. 2005). In its lower sequence, no lavas are interbedded with sedimentary deposits, and it is covered by a series of basalt flows and sills over 800 m thick with some thin beds of reddish-brown sandstone and claystone interbedded that constitute the intermediate sequence. The basalts have the typical characteristics of alkali–olivine basalts, commonly found in rift basins. Cuttings were dated by the K–Ar method and yielded 110 ± 6 Ma, indicating that the basalts were emplaced during Aptian times. Their composition closely resembles the rift-related basalts of Sierra Chica rather than those of the Paraná Magmatic Province (Webster et al. 2004; Chebli et al. 2005). In the sedimentary, lower section at least ten sills were drilled. These black intrusive rocks are mostly of basaltic nature, and their absolute K–Ar dating is widely divergent (Denison 1996 in Chebli et al. 2005) with an average of 103.3 Ma. Nevertheless, a dating obtained in a microdioritic sill was 147 ± 7 Ma (Oxfordian/Kimmeridgian times). It should be noted that this is significantly older than the K/Ar age of 110 Ma of the basalts overlying this lower sequence, and it is also difficult to reconcile with reliable biostratigraphic data recovered from this part of the column (Calegari et al. 2014). The most logical explanation is that an excess of radiogenic argon must have caused such an old age. It is worthy to note that perhaps the Levalle basin is buried because it is further south-east to Nazca flat slab (see Fig. 1 of Giménez et al. 2011) and, therefore, it was not seriously affected by Tertiary tectonics. As a consequence, rocks of the diverse sequences were not eroded. Despite having only one well that was drilled, it was possible to evaluate the huge thickness of about 800 m of volcanic rocks that form the intermediate sequence filling this basin.

2.2   Early Cretaceous in San Luis Province

To the west, in Sierra de las Quijadas and Cerrillada de las Cabras (Figs. 2.1 and 2.2a), Province of San Luis, Cretaceous basaltic rocks are recognized as lava flows and dykes interbedded with a sedimentary sequence that belong to the Las Salinas depocenter of the Western System Rift (Fig. 2.2a), and are comparable to those outcropping in Sierra Chica de Córdoba (Gordillo 1972; Costa et al. 2001). Available K–Ar dating indicates that ages range between 165 ± 3 Ma and 101 ± Ma (González 1971; Yrigoyen 1975; INGEIS 1977; Linares and González 1990). Their geochemical features are similar to those of the basalts of Córdoba, particularly to those of El Pungo locality (Martínez et al. 2012). Anyway, comparison between samples of Sierra Chica and those presented by Martínez et al. (2012) reveals lower contents in all the REE for the latter, particularly in the rock from Sierra de las Quijadas (Fig. 2.24a). In the same way, the multi-elemental diagram shows lower contents of Th, Nb, Ta and P, also specially in that sample (Fig. 2.24b), that lets infer differences in their respective mantle sources.

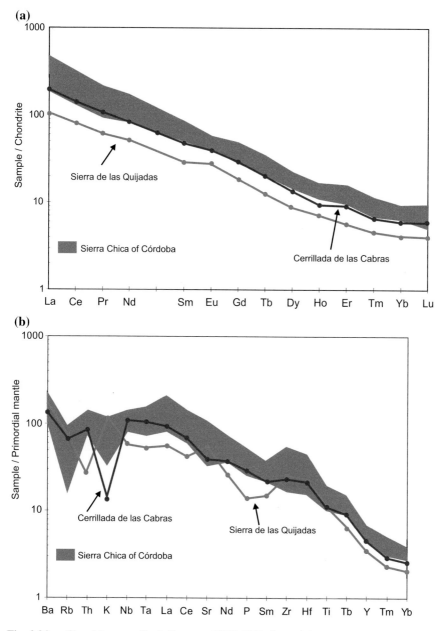

Fig. 2.24 **a** Chondrite-normalized (Boynton 1984) REE diagram for rocks of Sierra Chica of Córdoba (Lagorio 2008) in comparison with samples of Sierra de las Quijadas and Cerrillada de las Cabras from San Luis Province (Martínez et al. 2012). **b** Primordial mantle (Sun and McDonouth 1989) normalized multi-elemental plots for both groups of rocks, according those authors respectively

2.3   Probable Early Cretaceous Rocks in La Pampa and Buenos Aires Provinces

Further south, between the basins of Macachín–Quehué (Fig. 2.2a) and Colorado (La Pampa and Buenos Aires provinces; Fig. 2.2a), few and small outcrops of basalts were recognized by Silva Nieto et al. (2014). Their petrographic features seem to be more consistent with a subalkaline nature, instead of the alkaline one that characterizes volcanism in rifts of both Central and Western Systems. Nevertheless, geochemical data and radiometric ages are necessary for a better understanding of these volcanic rocks.

In the same way, basaltic levels have been reported (e.g. Raggio et al. 2012) in the subsurface of the aulacogenic Salado basin (Fig. 1.1b). Both geochemical and geochronological data are needed to characterize these rocks in the context of Gondwana break-up and the opening of the South Atlantic Ocean.

References

Ancheta MD, Sánchez ML, Marclé R (2002) Petrografía y geoquímica de las volcanitas de la Formación El Saucecito (Cretácico Inferior), Córdoba, Argentina. 15° Congreso Geológico Argentino, Actas, El Calafate, vol 2, pp 158–163

Astini RA, del Valle Oviedo N (2014) Cubierta sedimentaria mesozoica. In: Martino RD, Guereschi AB (eds) Geología y Recursos Naturales de la Provincia de Córdoba, Relatorio del 19° Congreso Geológico Argentino. Asociación Geológica Argentina, Córdoba, pp 435–472

Astini RA, Pezzi LI, Massei GA (1993) Paleogeografía y paleoambientes del Cretácico de la sierra de Pajarillo-Copacabana-Masa, noroeste de Córdoba. 12° Congreso Geológico Argentino, Actas, Mendoza, vol 1, pp 107–176

Bain Larrahona HG (1940) Estudios geológicos en la provincia de Córdoba. Boletín de Informaciones Petroleras. YPF, 192 p

Bellieni G, Piccirillo EM, Zanettin B (1981) Classification and nomenclature of basalts. IUGS, Subcommission of the Systematics of Igneous Rocks, Circular 34. Contrib Miner Petrol 87:1–19

Bellieni G, Comin-Chiaramonti P, Marques LS, Melfi AJ, Piccirillo EM, Nardy AJ, Roisemberg A (1984) High- and low-TiO_2 flood basalts from the Paraná plateau (Brasil): petrology and geochemical aspects bearing on their mantle origin. Neues Jahrb Mineral Abh 150:273–306

Bodenbender G (1907) Contribución al conocimiento geológico de la República Argentina. Anales del Ministerio de Agricultura y Secretaría de Geología, Mineralogía y Minería, Buenos Aires, vol 2(3), pp 1–35

Bodenbender G (1929) Triásico y Terciario de la falda oriental de las Sierras de Córdoba. Bol Acad Nac Cienc Córdoba 31:73–139

Boynton WV (1984) Cosmochemistry of the rare earth elements: meteorite studies. In: Henderson P (ed) Rare Earth element geochemistry. Elsevier, Amsterdam, pp 63–114

Bristow, JK (1984) Picritic rocks of the North Lebombo and South-East Zimbabwe. Geological Society of South-Africa Special Publication 13:105–123

Busby-Spera CJ, White JDL (1987) Variation in peperite textures associated with differing host-sediment properties. Bull Volc 49:765–775

Calegari RJ, Chebli G, Manoni RS, Lázzari V (2014) Las cuencas cretácicas de la región central del país: General Levalle. In: Martino RD, Guereschi AB (eds) Geología y Recursos Naturales de la Provincia de Córdoba, Relatorio del 19º Congreso Geológico Argentino. Asociación Geológica Argentina, Córdoba, 913–938

Caminos R, González P (1996) Mapa Geológico de la República Argentina. Escala 1:5.000.000. Dirección Nacional del Servicio Geológico. Secretaría de Minería de la Nación. Buenos Aires

Carlson RW, Esperança S, Svisero DP (1996) Chemical and Os isotopic study of Cretaceous potassic rocks from Southern Brazil. Contrib Miner Petrol 125:393–405

Carlson RW, Araujo ALN, Junqueira-Brod TC, Gasparf JC, Brod JA, Petrinovic IA, Hollanda MH, Pimentel MM, Sichel S (2007) Chemical and Isotopic relationships between peridotite xenoliths and magic-ultrapotassic rocks from Southern Brazil. Chem Geol 242:415–434

Carmichael ISE (1967) The iron-titanium oxides of salic volcanic rocks and their associated ferromagnesian silicates. Contrib Miner Petrol 14:36–63

Caroff M, Maury RC, Leterrier J, Joron JL, Cotten J, Guille G (1993) Trace element behaviour in the alkali basalt-comenditic trachyte series from Mururoa Atoll, French Polynesia. Lithos 30:1–22

Casquet C, Rapela CW, Pankhurst RJ, Baldo EG, Galindo C, Fanning CM, Dahlquist JA, Saavedra J (2012) A history of Proterozoic s in southern South America: from Rodinia to Gondwana. Geosci Front 3(2):137–145

Cejudo Ruiz R, Goguitchaivili A, Geuna SE, Alva-Valdivia L, Solé J, Morales J (2006) Early Cretaceous absolute geomagnetic paleointensities from Córdoba province (Argentina). Earth Planets Space 58(10):1333–1339

Chebli GA, Spalletti LA, Rivarola D, de Elorriaga E, Webster R (2005) Cuencas Cretácicas de la Región Central de la Argentina. Frontera Exploratoria de la Argentina. In: Chebli GA, Coriñas JS, Spalletti LA, Legarreta L, Vallejo EL (eds) 6º Congreso de Exploración y Desarrollo de Hidrocarburos, IAPG, Buenos Aires, pp 193–215

Comin-Chiaramonti P, Cundari A, Piccirillo EM, Gomes CB, Castorina F, Censi P, De Min A, Marzoli A, Speziale S, Velázquez VF (1997) Potassic and sodic igneous rocks from Eastern Paraguay: their origin from the lithospheric mantle and genetic relationships with the associated Paraná flood tholeiites. J Petrol 38:495–528

Comin-Chiaramonti P, Cundari A, DeGraff JM, Gomes CB and Piccirillo EM (1999) Early Cretaceous-Tertiary magmatism in Eastern Paraguay (western Paraná basin): geological, geophysical and geochemical relationships. J Geodyn 28:375–391

Comin-Chiaramonti P, Marzoli A, Gomes CB, Milan A, Riccomini C, Velásquez VF, Mantovani MSM, Renne P, Tassinari CCG, Vasconcelos PM (2007) Origin of Post-Paleozoic magmatism in Eastern Paraguay. In: Foulguer GR, Jurdy DM (eds) Plates, plumes, and planetary processes. Geological Society of America Special Papers, Boulder, Colorado, vol 430, pp 603–633

Comin-Chiaramonti P, De Min A, Cundari A, Girardi VAV, Ernesto M, Gomes CB, Riccomini C (2013) Magmatism in the Asunción-Sapucai-Villarrica Graben (Eastern Paraguay) Revisited: petrological, geophysical, geochemical and geodynamic inferences. Hindawi Publishing Corporation. J Geol Res 2013, Article ID 590835, 22 pp. 10.1155/2013/590835

Cortelezzi CR, Traversa L, Paulicevic PE (1981) Estudio petrológico y ensayos físicos de las rocas alcalinas del sur de las provincias de Córdoba y San Luis. 8º Congreso Geológico Argentino, Actas, San Luis, vol 4, pp 885–901

Costa CH, Gardini CE, Chiesa JO, Ortiz Suárez AE, Ojeda GE, Rivarola DL, Tognelli GC, Strasser EN, Carugno Durán AO, Morla PN, Guerstein PG, Sales DA, Vinciguerra HM (2001) Hoja Geológica 3366-III, San Luis. Provincias de San Luis y Mendoza, vol 293. Instituto de Geología y Recursos Minerales, Servicio Geológico Minero Argentino, Boletín, Buenos Aires, 67 p

Cox KG (1988) The Karoo province. In: MacDougall JD (ed) Continental flood basalts. Kluwer Academic, Dordrecht, pp 239–271

Cundari A, Ferguson AK (1982) Significance of the pyroxene chemistry from leucite bearing and related assemblages. Tschermaks Mineralogische und Petrographische Mitteilungen 30:189–204

Cundari A, Comin-Chiaramonti P (1996) Mineral chemistry of alkaline rocks from the Asunción–Sapukai graben (central-eastern Paraguay). In: Comin-Chiaramonti P, Gomes CB (eds.) Alkaline magmatism in central-eastern Paraguay. Relationships with coeval magmatism in Brazil. Edusp/Fapesp, São Paulo, pp 181–194

De la Roche H, Leterrier P, Grandclaude P, Marchal M (1980) A classification of volcanic and plutonic rocks using R1-R2 diagram and major element analysis. Its relationships with current nomenclature. Chem Geol 29:183–210

De Min A (1993) Il magmatismo mesozoico K-alcalino del Paraguay Oriental: aspetti petrogenetici ed implicanze geodinamiche. Unpublished doctoral thesis, Universitá di Trieste, Trieste, 242 p

De Paolo DJ (1981) Trace elements and isotopic effects of combined wallrock assimilation and fractional crystallization. Earth Planet Sci Lett 43:201–211

Deer WA, Howie RA, Zussman J (1992) An introduction to the rock-forming minerals. Longman Scientific and Technical, Essex, 696 p

Delpino D, Sánchez ML, Bermúdez A, Marclé R, Ancheta D (1999) Los depósitos hidroclásticos y estrombolianos del Grupo El Pungo, Córdoba. 14° Congreso Geológico Argentino, Actas, Salta, vol 2, pp 197–199

Escayola MP, Viramonte JG, Becchio R, Franz G, Arnosio M, Popridkin MC (1998) Xenolitos en volcanitas alcalinas cretácicas del sector sur de la Sierra de Los Cóndores, Sierras Pampeanas de Córdoba, Argentina. 10° Congreso Latinoamericano de Geología y 6° Congreso Nacional de Geología Económica, Actas, Buenos Aires, vol 2, pp 354–358

Escayola MP, Franz G, Becchio R, Viramonte JG, Lucassen F (1999) Naturaleza y evolución de la litósfera subcontinental durante el rifting Cretácico en las Sierras Pampeanas de Córdoba, Argentina. 14° Congreso Geológico Argentino, Actas, Salta, vol 2, pp 200–203

Escayola MP, Pimentel MM, Armstrong R (2007) Neoproterozoic backarc basin: Sensitive high-resolution ion microprobe U-Pb and Sm-Nd isotopic evidence from the Eastern Pampean Ranges, Argentina. Geology 35(6):495–498

Ferreira Pittau ML, Escayola MP, Garrido A, (2008) Ciclos eruptivos del Grupo Sierra de los Cóndores (Sierra Chica de Córdoba): Nuevas características petrológicas y evolutivas. 17° Congreso Geológico Argentino, Actas, vol 3, pp 1347–1348

Ferreira L, Escayola MP, Viramonte JG, Franz G (1999) Secuencias volcano-sedimentarias del complejo volcánico Los Cóndores, Córdoba: litoestratigrafía y mecanismos eruptivos. 14° Congreso Geológico Argentino, Actas, Salta, vol 2, pp 204–206

Galliski MA, Dorais M, Lira R (1996) Las pegmatitas ijolíticas de La Madera, provincia de Córdoba: Quimismo de sus minerales y model genético. 13° Congreso Geológico Argentino y 3° Congreso de Exploración de Hidrocarburos, Actas, vol 3, pp 207–225.

Galliski MA, Lira R, Dorais MJ (2004) Low-pressure differentiation of melanephelinitic magma and the origin of ijolite pegmatites at La Madera, Córdoba, Argentina. The Canadian Mineralogist 42(6):1799–1823

Geuna S (1997) Geología y paleomagnetismo de unidades cretácicas de la provincia de Córdoba. Unpublished doctoral thesis, Universidad Nacional de Córdoba, Córdoba, 263 p

Geuna SE (1998) Paleomagnetismo del Grupo Sierra de Los Cóndores (Cretácico Inferior de Córdoba): correlación magnetoestratigráfica local, y sus consecuencias en la interpretación geológica. Rev Asoc Geol Argentina 53(1):69–82

Geuna SE, Vizán H (1998) New Early Cretaceous palaeomagnetic pole from Córdoba Province (Argentina): revision of previous studies and implications for the South American database. Geophys J Int 135:1085–1100

Geuna SE, Escosteguy LD, Miró R, Candiani JC, Gaido MF (2008) La susceptibilidad magnética del Batolito de Achala (Devónico, Sierra Grande de Córdoba) y sus diferencias con otros granitos achalianos. Rev Asoc Geol Argentina 63(3):380–394

Geuna SE, Lagorio SL, Vizan H (2015) Oxidation processes and their effects on the magnetic remanence of Early Cretaceous subaerial basalts from Sierra Chica de Córdoba, Argentina. In: Ort MH, Porreca M, Geissman JW (eds) The use of palaeomagnetism and rock magnetism to understand volcanic processes, vol 396. Geological Society Special Publication, London, pp 239–263

Gibson SA, Thompson RN, Leonardos OH, Dickin AP, Mitchell JG (1995) The late cretaceous impact of the Trinidade mantle plume: evidence from large-volume, mafic, potassic magmatism in SE Brasil. J Petrol 36:189–229

Gibson SA, Thompson RN, Dickin AP, Leonardos OH (1996) Erratum to "High-Ti and low-Ti mafic potassic magmas: key to plume-litosphere interactions and continental flood-basalt genesis". Earth Planet Sci Lett 141:325–341

Gibson SA, Thompson RN, Day JA (2006) Timescales and mechanism of plume lithosphere interactions: $^{40}Ar/^{39}Ar$ geochronology and geochemistry of alkaline igneous rocks from the Paraná-Etendeka large igneous province. Earth Planet Sci Lett 251:1–17

Giménez ME, Dávila F, Astini R, Martínez P (2011) Interpretación gravimétrica y estructura cortical en la Cuenca de General Levalle, Provincia de Córdoba, Argentina. Rev Mex Cienc Geol 28(1):105–117

González RR (1971) Edades radimétricas de algunos cuerpos eruptivos de la República Argentina. Rev Asoc Geol Argentina 26(3):411–412

González RR, Kawashita K (1972) Edades potasio-argón de rocas básicas de la provincia de Córdoba. Rev Asoc Geol Argentina 27:259–260

Gordillo CE (1972) Petrografía y composición química de los basaltos de la sierra de Las Quijadas – San Luis – y sus relaciones con los basaltos cretácicos de Córdoba, vol 1(3–4). Boletín de la Asociación Geológica de Córdoba, Córdoba, pp 127–129

Gordillo CE, Lencinas A (1967a) Geología y petrología del extremo norte de la Sierra de Los Cóndores, Córdoba, vol 46(1). Boletín Academia Nacional de Ciencias, Córdoba, pp 73–108

Gordillo CE, Lencinas A (1967b) El basalto nefelínico de El Pungo, Córdoba, vol 46(1). Boletín Academia Nacional de Ciencias, Córdoba, pp 109–115

Gordillo CE, Lencinas A (1969) Perfil geológico de la sierra Chica de Córdoba en la zona del río Los Molinos, con especial referencia a los diques traquibasálticos que la atraviesan, vol 47. Boletín Academia Nacional de Ciencias, Córdoba, pp 27–50

Gordillo CE, Lencinas A (1980) Sierras Pampeanas de Córdoba y San Luis. In: Turner JCM (ed) 2° Simposio de Geologia Regional Argentina 1, Academia Nacional de Ciencias, Córdoba, pp 577–650

Gordillo CE, Linares E, Daziano C (1983) Nuevo afloramiento de nefelinita olivínica: Estancia Guasta, Sierras de Córdoba. Rev Asoc Geol Argentina 28 (3–4): 485–489

Green DH (1970) A review of experimental evidence on the origin of basaltic and nephelinitic magmas. Phys Earth Planet Inter 3:221–235

Hanson GN (1978) The application of trace elements to the petrogenesis of igneous rocks of granitic composition. Earth Planet Sci Lett 38:26–43

Iannizzotto NF, Rapela CW, Baldo EGA, Galindo C, Fanning CM, Pankhurst RJ (2013) The Sierra Norte-Ambargasta batholiths: late Ediacaran-Early Cambrian magmatism associated with Pampean transpressional tectonics. J S Am Earth Sci 42:127–143

INGEIS (1977) Nuevas constantes a utilizar en los métodos de datación radimétrica. Rev Asoc Geol Argentina 32:239–240

Kay SM, Ramos VA (1996) El magmatismo cretácico de las sierras de Córdoba y sus implicancias tectónicas. 13° Congreso Geológico Argentino y 3° Congreso de Exploración de Hidrocarburos, Actas, Buenos Aires, vol 3, pp 453–464

Kraemer PE, Escayola MP, Martino RD (1995) Hipótesis sobre la evolución tectónica neoproterozoica de las Sierras Pampeanas de Córdoba (30° 40'–32° 40'), Argentina. Rev Asoc Geol Argentina 50(1–4):47–59

Kuiper KF, Deino A, Hilgen FJ, Krijgsman W, Renne PR, Wijbrans JB (2008) Synchronizing rock clocks of Earth history. Science 320:500–504

Kull V, Methol E (1979) Descripción geológica de la Hoja 21i, Alta Gracia. Dirección Nacional de Geología y Minería. Boletín, Buenos Aires, vol 55, p 72

Lagorio SL (1998) Geoquímica y petrogénesis de volcanitas cretácicas de la sierra Chica de
    Córdoba (Argentina). 10° Congreso Latinoamericano de Geología y 6° Congreso Nacional de
    Geología Económica, Actas, Buenos Aires, vol 2, pp 314–320
Lagorio SL (2003) El volcanismo cretácico alcalino de la sierra Chica de Córdoba: geoquímica,
    petrogénesis e implicancias geodinámicas. Unpublished doctoral thesis, Universidad Nacional
    de Buenos Aires, Buenos Aires, 400 p
Lagorio SL (2008) Early Cretaceous alkaline volcanism of the Sierra Chica de Córdoba
    (Argentina): mineralogy, geochemistry and petrogenesis. J S Am Earth Sci 26:152–171
Lagorio SL, Montenegro TF (2004) Nódulos lherzolíticos espinélicos en basaltos alcalinos del
    norte de la sierra de los Cóndores (Córdoba). In: Brodtkorb M, Koukharsky M, Quenardelle S,
    Montenegro T (eds) Avances en Mineralogía, Metalogenia y Petrología 2004. Universidad de
    Buenos Aires y Universidad Nacional de Río Cuarto, pp 343–348
Lagorio SL, Geuna SE, Iacumin M, Vizán H (1997) Características geoquímicas del volcanismo
    cretácico del sector norte de la Sierra de Los Cóndores (Córdoba, Argentina). 8° Congreso
    Geológico Chileno, Actas, Antofagasta, vol 2, 1334–1338
Lagorio SL, Vizán H, Geuna SE (2014) El volcanismo alcalino cretácico. In: Martino RD,
    Guereschi AB (eds) Geología y Recursos Naturales de la Provincia de Córdoba, Relatorio del
    19° Congreso Geológico Argentino. Asociación Geológica Argentina, Córdoba, pp 473–511
Le Bas MJ, Le Maitre RW, Strekeisen A, Zanetin B (1986) A chemical classification of volcanic
    rock based on the total alkali-silica diagram. J Petrol 27:745–750
le Roex AP (1985) Geochemistry, mineralogy and magmatic evolution of the basaltic and trachytic
    lavas form Gough Island, South Atlantic. J Petrol 26:149–186
le Roex A, Cliff RA, Adair BJL (1990) Trsitan da Cunha, South Atlantic: geochemistry and
    petrogeneis of a basanite-phonolite lavas series. J Petrol 31:779–812
Lencinas AN (1971) Geología del valle de Punilla entre Bialet Masse y La Cumbre, provincia de
    Córdoba, vol 1(2). Boletín Asociación Geológica de Córdoba, Córdoba, pp 61–70
Linares E, González R (1990) Catálogo de edades radimétricas de la República Argentina 1957–
    1987. Publicaciones especiales de la Asociación Geológica Argentina, Serie B, Didáctica y
    Complementaria, vol 19. Ascociación Geológica Argentina, Buenos Aires, 628 p
Linares E, Valencio DA (1974) Edades Potasio-Argón y paleomagnetismo de los diques
    traquibasálticos del río de Los Molinos, Córdoba, República Argentina. Rev Asoc Geol
    Argentina 29(3):341–348
Lira R, Sfragulla J (2014) El magmatismo devónico-carbonífero: El batolito de Achala y los
    plutones menores al norte del cerro Champaquí. In: Martino RD, Guereschi AB (eds) Geología
    y Recursos Naturales de la Provincia de Córdoba, Relatorio del 19° Congreso Geológico
    Argentino. Asociación Geológica Argentina, Córdoba, pp 293–347
Lira R, Poklepovic MF, O'Leary MS (2014) El magmatismo cámbrico en el batolito de Sierra
    Norte-Ambargasta. In: Martino RD, Guereschi AB (eds) Geología y Recursos Naturales de la
    Provinica de Córdoba, Relatorio del 19° Congreso Geológico Argentino. Asociación Geológica
    Argentina, Córdoba, pp 183–215
Lucassen F, Escayola MP, Romer RL, Viramonte JG, Koch K, Franz G (2002) Isotopic
    composition of Late Mesozoic basic and ultrabasic rocks from the Andes (23–32° S)—
    implications for the Andean mantle. Contrib Miner Petrol 143:336–349
Lucassen F, Franz G, Romer RL, Schultz F, Dulski P, Wemmer K (2007) Pre-Cenozoic intra-plate
    magmatism along the Central Andes (17–34°S): Composition of the mantle at an active margin.
    Lithos 99: 312–338
Lucero Michaut NH, Gamkosian A, Jarsun B, Zamora E, Sigismondi M, Miró R, Caminos R
    (1995) Mapa Geológico de la Provincia de Córdoba. Secretaría de Minería, Dirección Nacional
    del Servicio Geológico
Martino R, Kraemer P, Escayola MP, Giambastiani M, Arnosio M (1995) Transecta de las Sierras
    Pampeanas de Córdoba a los 32° S. Rev Asoc Geol Argentina 50(1–4):60–77
Martino RD, Guereschi AB, Carignano CA, Calegari R, Manoni R (2014) La estructura de las
    cuencas extensionales cretácicas de las Sierras de Córdoba. In: Martino RD, Guereschi AB

(eds) Geología y Recursos Naturales de la Provincia de Córdoba, Relatorio del 19° Congreso Geológico Argentino. Asociación Geológica Argentina, Córdoba, pp 513–538

Marzoli A, Melluso L, Morra V, Renne PR, Sgrosso I, D'Antonio M, Duarte Morais L, Morais EAA, Ricci G (1999) Geochronology and petrology of Cretaceous basaltic magmatism in the Kwanza basin (western Angola), and relationships with the Paraná-Etendeka continental flood basalt province. J Geodyn 28:341–356

MacGregor ID (1974) The system $MgO–Al_2O_3–SiO_2$: solubility of Al2O3 in enstatite for spinel and garnet peridotite compositions. American Mineralogist 59:110–119

Mendía JE (1978) Paleomagnetic study of alkaline vulcanites from Almafuerte, province of Córdoba, Argentina. Geophys J Roy Astron Soc 54:539–546

Minudri CA, Sánchez ML (1994) Paleoambientes de sedimentación de la sección superior del Grupo Sierra de Los Cóndores (Cretácico Inferior), Córdoba, Argentina. 5° Reunión Argentina de Sedimentología, Actas, pp 29–34

Morimoto K, Fabries J, Ferguson AK, Ginzburg IV, Ross M, Seifert FA, Zussman J, Aoki K, Gottardi G (1988) Nomenclature of pyroxenes. Mineral Petrol 39:55–76

Nimis P (1995) A clinopyroxene geobarometer for basaltic systems based on crystal-structure modeling. Contrib Miner Petrol 121:115–125

Nimis P, Ulmer P (1998) Clinopyroxene geobarometry of magmatic rocks. Part 1: an expanded structural geobarometer for anhydrous and hydrous, basic and ultrabasic systems. Contrib Miner Petrol 133:122–135

O'Brien HE, Irving AI, McCallum IS (1988) Complex zoning and resorption of phenocrysts in mixed potassic mafic magmas of the Highwood Mountains, Montana. Am Mineral 73:1007–1024

O'Hara MJ, Mathews RE (1981) Geochemical evolution in an advancing, periodically replenished, periodically tapped, continuously fractionated magma chamber. J Geol Soc London 138:237–277

O'Leary MS, Lira R, Poklepovic MF (2014) Volcanismo y subvolcanismo del sector centro-oeste del batolito Sierra Norte-Ambargasta. In: Martino RD, Guereschi AB (eds) Geología y Recursos Naturales de la Provincia de Córdoba, Relatorio 19° Congreso Geológico Argentino. Asociación Geológica Argentina, Córdoba, pp 217-232

Oviedo N, Astini RA (2014) Depósitos volcanosedimentarios del Cretácico en Las Cumbres y vertiente oriental de la Sierra Chica, Córdoba, y redefinición de la Formación El Pungo. Rev Asoc Geol Argentina 71(4):472–483

Papike JJ, Cameron K, Baldwin K (1974) Amphiboles and pyroxenes: characterization of other than quadrilateral components and estimates of ferric iron from microprobe data. Bull Geol Soc Am 6:1053–1054

Pastore F (1930) Notas sobre Triasico y Terciario de la falda oriental de las Sierras de Cordoba. Relaciones morfológico-tectónicas. Rocas volcánicas del doctor Bodenbender. Anales de la Sociedad Cientifica Argentina 110:399–407

Peate DW (1997) The Paraná-Etendeka Province. In: Mahoney JJ, Coffin MF (eds) Large igneous provinces: continental oceanic and planetary flood volcanism, vol 100. Geophysical Monograph American Geophysical Union, Boulder, Colorado, pp 215–245

Peate DW, Hawkesworth CJ (1996) Lithospheric to asthenospheric transition in low-Ti flood basalts from Southern Paraná, Brazil. Chem Geol 127:1–24

Pensa M (1957) Contribución al conocimiento de los meláfiros en las sierras de Córdoba. Representación cartográfica y perfiles. Relaciones morfológicas, tectónicas y correlación con las diversas efusiones en el país. Revista Universidad Nacional de Córdoba, Facultad de Ciencias Exactas, Físicas y Naturales, vol 19(3–4), pp 471–500

Philpotts AR (1976) Silicate liquid immiscibility: its probable extent and petrogenetic significance. Am J Sci 276:1147–1177

Piccirillo EM, Melfi AJ (1988) The Mesozoic Flood Volcanism from the Paraná Basin (Brazil): petrogenetic and geophysical aspects. Universidade de São Paulo, San Pablo, 600 p

Piovano EL (1994) Facies de mantos de crecida y cauces efímeros en la Formación Saldán, Cretácico Inferior, sierra Chica de Córdoba. 5° Reunión Argentina de Sedimentología, Actas, pp 35–40

Piovano EL (1996) Correlación de la Formación Saldán (Cretácico temprano) con otras secuencias de las Sierras Pampeanas y de las cuencas Chacoparanense y de Paraná. Rev Asoc Geol Argentina 51(1):29–36

Piovano EL, Astini RA (1990) Facies de abanico aluvial semiárido en la Formación Saldán, quebrada del río Suquía, Sierra Chica de Córdoba. 3° Reunión Argentina de Sedimentología, Actas, San Juan, pp 217–222

Poiré D, Sánchez ML, Villegas M (1989) Facies sedimentarias de la sección inferior del Grupo Sierra de Los Cóndores, Embalse Río Tercero, Provincia de Córdoba, República Argentina. Contribuciones de los Simposios sobre Cretácico de América Latina. Parte A: Eventos y Registro Sedimentario, pp 121–132

Quenardelle S, Montenegro TF (1998) Las rocas fóidicas de Córdoba (Chaján) y San Luis (Las Chacras), Argentina. Petrología y Geoquímica. 10° Congreso Latinoamericano de Geología y 6° Congreso Nacional de Geología Económica, Actas, Buenos Aires, vol 2, pp 300–305

Ramos VA (1999) Evolución tectónica de la Argentina. Geología Argentina. Anales del Instituto de Geología y Recursos Minerales 29(24):715–784

Ramos VA, Escayola MP, Mutti DI, Vujovich GI (2000) Proterozoic-early Paleozoic ophioltes of the Andean basement of southern South America. In: Dilek Y, Moores EM, Elthon D, Nicholas A (eds) Ophiolites and Oceanic Crust: new insights from field studies and the Ocean Drilling Program, Geological Society of America Special Paper, Boulder, Colorado, vol 349, 331–349

Raggio F, Gerster R, Welsink H (2012) Cuencas del Salado y Punta del Este. Petrotecnia, diciembre 2012

Rapela CW, Pankhurst RJ, Casquet C, Baldo E, Saavedra J, Galindo C, Fanning CM (1998) The Pampean Orogeny of the southern proto-Andes: cambrian continental collision in the Sierras de Córdoba. In: Pankhurst RJ, Rapela CW (eds) The Proto-Andean Margin of Gondwana, vol 142. Geological Society London Special Publications, pp 181–217

Rapela CW, Pankhurst RJ, Casquet C, Fanning CM, Baldo EG, González-Casado JM, Galindo C, Dahlquist J (2007) The Río de la Plata craton and the assembly of SW Gondwana. Earth-Sci Rev 83:49–82

Rapela CW, Fanning CM, Casquet C, Pankhurst RJ, Spalletti L, Poiré D, Baldo EG (2011) The Río de la Plata craton and the adjoining Pan-African/brasiliano shield: their origins and incorporation into south-west Gondwana. Gondwana Res 20(4):673–690

Ringwood AE (1966) The chemical composition and origin of the Earth. In: Hurley PM (ed) Advances in earth sciences: contribution to the international conference on the earth sciences. Cambridge MIT Press, pp 287–356

Rossello E, Mozetic ME (1999) Caracterización estructural y significado geotectónico de los depocentros cretácicos continentales del centro–oeste argentino. Boletim do 5° Simpósio sobre o Cretáceo do Brasil, UNESP–Campus de Rio Claro/SP:107–113

Rudnick RL, Gao S (2003) Composition of the continental crust. In: Rudnick RL (ed) Tretease on geochemistry, vol 3. The Crust. Elsevier, Amsterdam, p 1–64

Sánchez ML (2001a) Sedimentología de la Formación El Rosario (Cretácico), La Cumbre, provincia de Córdoba, Argentina. 11° Congreso Latinoamericano de Geología y 3° Congreso Uruguayo de Geología, versión CD-ROM, Montevideo

Sánchez ML (2001b) Sedimentología de la Formación Peñón Blanco (Cretácico), en la zona de La Cumbre, provincia de Córdoba, Argentina. 11° Congreso Latinoamericano de Geología y 3° Congreso Uruguayo de Geología, CD-ROM, Montevideo

Sánchez ML, Bermúdez A (1997) Caracterización geoquímica del volcanismo cretácico de la Sierra de los Cóndores, Córdoba. Argentina. 8° Congreso Geológico Chileno, Actas, Antofagasta, vol 2, pp 1522–1527

Sánchez ML, Villegas MB, Poiré DG (1990) Paleogeografía del Cretácico Inferior en el área de la sierra de Los Cóndores, provincia de Córdoba. 3° Reunión Argentina de Sedimentología, Actas, San Juan, pp 235–246

Sánchez ML, Pérez Posio C, Sisto F (1995) Modelo cinemático de la cuenca de Los Cóndores (Cretácico Inferior) en la provincia de Córdoba. 5° Jornadas Pampeanas de Ciencias Naturales, Actas, Santa Rosa, 1993, vol 2, pp 75–81

Sánchez ML, Toro E, Delpino D, Bermúdez A (1999) Geología y estratigrafía de las rocas cretácicas de La Cumbre - Estancia El Rosario, Córdoba. 14° Congreso Geológico Argentino, Actas, Salta, vol 1, pp 445–448

Sánchez ML, Ancheta MD, Marclé R (2001) Las volcanitas de la Formación El Saucecito, Grupo El Pungo (Cretácico Inferior), Córdoba, Argentina. 11° Congreso Latinoamericano de Geología y 3° Congreso Uruguayo de Geología, CD-ROM, Montevideo

Sánchez ML, Ancheta MD, Marcle R (2002) Rocas volcánicas cretácicas de la cuenca de El Pungo, Córdoba. 15° Congreso Geológico Argentino, Actas CD-ROM, El Calafate

Schmidt CJ, Astini RA, Costa CH, Gardini CE, Kraemer PE (1995) Cretaceous rifting, alluvial fan sedimentation and Neogene inversion, southern Sierras Pampeanas, Argentina. In: Tankard AJ, Suárez Soruco R, Welsink HJ (eds) Petroleum basins of South America, vol 62. American Association of Petroleum Geologists, Memoir, Tulsa, pp 341–357

Schröder C (1967) Estudio geológico y geotécnico referente al aprovechamiento hidroeléctrico y para riego del río Tercero en "Piedras Moras". Prov. de Córdoba. Trabajo Final de Licenciatura, Universidad de Buenos Aires (unpublished), Buenos Aires 47 p

Shelley D (1993) Igneous and Metamorphic rocks under the microscope. Chapman and Hall, London, 445 p

Shimizu N, Le Roex AP (1986) The chemical zoning of augite phenocrysts in alkaline basalts from Gough Island, South Atlantic. J Volcanol Geoth Res 29:159–188

Silva Nieto DG, Espejo PM, Lagorio SL (2014) Hallazgo de basaltos de probable edad cretácica en el centro este de La Pampa. 19° Congreso Geológico Argentino, Actas CD-Rom, Córdoba

Sisto FA, Cortés JM (1992) Tectónica cretácico-cenozoica del tramo sur de la Sierra de Los Cóndores, Sierras Pampeanas de Córdoba. 7° Reunión de Microtectónica, Actas, pp 63–69

Sisto F, Sánchez ML, Perez Posio C (1993) Dinámica y cinemática cenozoica del frente de fracturación Puesto Viejo - La Cantera, sierra de los Cóndores, provincia de Córdoba, República Argentina. 9° Reunión de Microtectónica, Actas, Mendoza, pp 7–8

Sisto F, Sánchez ML, Perez Posio C, (1995) Dinámica y cinemática cenozoica de las principales estructuras de la Sierra de los Cóndores, provincia de Córdoba, República Argentina. 5° Jornadas Pampeanas de Ciencias Naturales, Actas, Santa Rosa, 1993, vol 2, pp 82–88

Spencer KJ, Lindsley DH (1981) A solution model for coexisting iron-titanium oxides. Am Mineral 66:1189–1201

Stipanicic PN, Linares E (1969) Edades radimétricas determinadas para la República Argentina y su significado geológico, vol 47(1). Boletín de la Academia Nacional de Ciencias, Córdoba, pp 51–96

Stormer JC, Nicholls J (1978) XLFRAC: A program for the interactive testing of magmatic testing of magmatic differentiation models. Comput Geosci 4:143–159

Sun SS, McDonough WF (1989) Chemical and isotopic systematics of oceanic basalts: Implications for mantle composition and processes. In: Saunders AD, Norry MJ (eds) Magmatism in the ocean basins, vol 42. Special Publication Geological Society of London, London, pp 313–345

Takahashi E, Kushiro I (1983) Melting of a dry peridotite at high pressure and basalt magma genesis. Am Mineral 68:859–879

Tankard AJ, Uliana MA, Welsnik HJ, Ramos VA, Turic M, França AB, Milani EJB, de Brito Neves B, Eyles N, Skarmeta J, Santa Ana H, Wiens F, Cirbián M, López Paulsen O, Germs GJB, De Wit MJ, Machacha T, McG. Miller R (1995) Structural and tectonic controls of basin evolution in Southwestern Gondwana during the Phanerozoic. In: Tankard AJ, Suárez Soruco R, Welsink HJ (eds) Petroleum basins of South America, vol 62. American Association of Petroleum Geologists, Memoir, Tulsa, pp 5–52

Thover E, Trindade RIF, Solum JG, Hall CM, Riccomini C, Nogueira AC (2010) Closing the
    Clymene ocean and bending a Brasiliano belt: evidence for the Cambrian formation of
    Gondwana, southeast Amazon craton. Geology 38(3):267–270
Uliana MA, Biddle KT, Cerdan J (1990) Mesozoic extension and the formation of Argentine
    sedimentary basins. In: Tankard AJ, Balkwill HR (eds) Extensional tectonics and stratigraphy
    of the North Atlantic margins, vol 46. American Association of Petroleum Geologists, Memoir,
    Tulsa, 599–614
Valencio DA (1972) Palaeomagnetism of the lower Cretaceous Vulcanitas Cerro Colorado
    Formation of the Sierra de los Cóndores Group, province of Córdoba, Argentina. Earth Planet
    Sci Lett 16:370–378
Valencio DA, Vilas JF (1972) Sequence of the continental movements occurred prior to and after
    the formation of the South Atlantic. Anais de la Academia Brasileira de Ciencias 48:377–386
Valencio DA, López MG, Solá P, Villani C (1980) El significado geológico de los resultados del
    estudio paleomagnético de vulcanitas alcalinas de las provincias de San Luis y Córdoba. Rev
    Asoc Geol Argentina 25(3):340–347
Viramonte J, Deruelle B, Moorbath S, Mazzuoli R, Omarini R (1994) El volanismo alcalino de
    Chaján–Las Chacras, Córdoba–San Luis, Argentina. 7° Congreso Geológico Chileno, Actas,
    Concepción, vol 2, 1273–1277
Webster RE, Chebli GA, Fischer FJ (2004) General Levalle basin, Argentina: a frontier Lower
    Cretaceous rift basin. Am Assoc Petrol Geol Bull 88(5): 627–652
Wilkinson JFG, Le Maitre RW (1987) Upper mantle amphiboles and micas and TiO_2, K_2O and
    P_2O_5 abundance and 100 $Mg/(Mg + Fe^{2+})$ ratios of common basalts and andesites: implications
    for modal mantle metasomatism and undepleted mantle compositions. Journal of Petrology
    28:37–73
Yrigoyen MR (1975). La edad cretácica del Grupo Gigante (San Luis) y su relación con cuencas
    circunvecinas. 1° Congreso Argentino de Paleontología y Bioestratigrafía, Actas, Tucumán, vol
    2, pp 29–56
Zindler A, Hart S (1986) Chemical geodynamics. Ann Earth Planet Sci 14:493–571

Chapter 3
Early Cretaceous Volcanism in Eastern Argentina

Abstract Volcanic rocks of Paraná Magmatic Province (PMP) are mostly located within the area of the intracratonic Paraná basin and covered an extensive region of Brazil, Paraguay, Uruguay and north-eastern Argentina. In the latter, the main outcrops are located in Misiones and Corrientes provinces, while a thick volcanic pile remains buried under the surface of Entre Ríos Province and in Chaco–Paraná basin (southern Paraná). The geochemical data presented here complement those published by other authors. In Misiones Province, rocks are tholeiitic lavas of high- and low-Ti varieties, whereas to the south, in Corrientes Province, lavas of low-Ti types prevail. Cu contents are high, comparable to those reported from lavas of Brazil. Evolution of HTi and LTi magmas must have occurred from two different parental magmas through fractional crystallization at low pressures, involving crustal contamination only for the Gramado variety, in agreement with the general trend recognized in the PMP. Geochemical features point out that different ancient subduction processes could have affected the source of the PMP basalts from Misiones Province, as those that occurred during Late Proterozoic–Early Cambrian associated to Paraguay belt, as well as even older ones related to the Nico Pérez terrane.

Keywords Tholeiitic basalts · Misiones · Corrientes · Paraná · High-Ti basalts · Low-Ti basalts

As has been mentioned previously, the volcanism of the Paraná Magmatic Province (PMP) covered an extensive region of Brazil, Argentina, Paraguay and Uruguay. In Argentina, lava flows poured out over the Mesopotamia region composed of the provinces of Misiones, Corrientes and Entre Ríos, as well as in the Chaco–Paraná basin (Fig. 3.1a). Its outcrops are presently restricted to Misiones and part of Corrientes provinces, as in the remaining areas lavas lay in the subsurface.

© The Author(s) 2016
S.L. Lagorio et al., *Early Cretaceous Volcanism in Central and Eastern Argentina During Gondwana Break-Up*, SpringerBriefs in Earth System Sciences, DOI 10.1007/978-3-319-29593-0_3

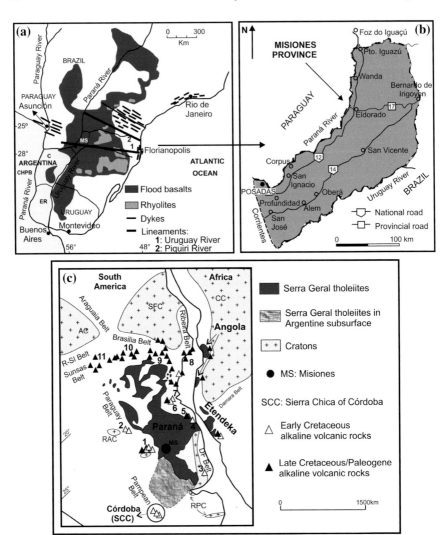

Fig. 3.1 a Schematic map of the outcrops of the continental flood basalts of Paraná Magmatic Province (PMP), adapted from Piccirillo and Melfi (1988); *MS* Misiones Province, *C* Corrientes Province, *ER* Entre Ríos Province, *CHPB* Chaco–Paraná basin. **b** Map of Misiones Province (MS), Argentina with location of sampled localities. Based on Lagorio and Vizán (2011) **c** Illustration of the Paraná–Etendeka–Angola Large Igneous Province (PEA LIP), at circa 130 Ma, including also the peripheral alkaline localities. Adapted from Piccirillo and Melfi (1988); Gibson et al. (1996); Marzoli et al. (1999); Lagorio (2008); Lagorio and Vizán (2011). *RPC* Río de la Plata craton, *SFC* San Francisco craton, *AC* Amazonian craton, *CC* Congo craton, *RAC* Río Apa craton, *R-SI* Rondonia–San Ignacio belt. *1* eastern Paraguay, *2* Amambay, *3* Mariscala, *4* Anitápolis, *5* Lages, *6* Ponta Grossa, *7* Jacupiranga, *8* Serra do Mar, *9* Alto Paranaíba, *10* Goiás, *11* Poxoreu

3.1 Early Cretaceous in Misiones Province

3.1.1 Geological Setting

Volcanic rocks of the Paraná Magmatic Province are mostly located within the area of the Eopaleozoic intracratonic Paraná basin. Valuable information on different aspects of these continental flood basalts may be found in the works of several authors (Piccirillo and Melfi 1988; Hawkesworth et al. 1992; Peate 1997; Marques et al. 1999; Ernesto et al. 1999; Gibson et al. 2006; Thiede and Vasconcelos 2010; Rocha-Junior et al. 2012, 2013, among others). This large igneous province covered an extensive region of Brazil, Paraguay, Uruguay and north-eastern Argentina (Figs. 1.1, 1.2 and 3.1a).

In Brazil, this volcanism is assigned to the Serra Geral Formation, whereas in Argentina to the Posadas Formation (Gentili and Rimoldi 1980), although the name Serra Geral is used as well. More recently, lavas from Brazil were considered as a Group, composed by several formations (Wildner et al. 2007). On the other hand, in Uruguay it is known as the Arapey Group, that also consists of various formations including interbedded sedimentary levels (see, e.g. Bossi and Schipilov 2007). Considering the whole magmatic province, it is represented by tholeiitic basalts and basaltic andesites (90 %), along with subordinate andesites (7 %) and rhyodacites–rhyolites (3 %), according to Piccirillo and Melfi (1988). From a petrology point of view, these authors divided the PMP into three main sections (Fig. 3.1a): (1) northern section, located north of the Piquiri River Lineament, (2) central section, between the latter and the Uruguay River Lineament and (3) the southern section.

3.1.1.1 Lava Flows of Misiones Province in the Context of the Paraná Magmatic Province

In Misiones Province, the basalt outcrops of the Serra Geral Formation mostly belong to the central section and to a lesser extent to the southern one of PMP (Fig. 3.1a). Lava flows are mainly subhorizontal (Fig. 3.2a) and show thickness ranging from a few metres to several hundreds of metres. It should be noted that more than 1200 m have been reported from a borehole at Oberá locality (Ardolino and Miranda 2008). Lava poured out over aeolian sands of the Solari Formation, whose contact is only exposed in the San Ignacio area (Fig. 3.1b; e.g. Marengo et al. 2005). Some breccias interbedded between the flood basalts are also observed (Fig. 3.2b, c). They were considered as peperitic levels, of pillowy type of Busby-Spera and White (1987), indicating an interaction between lavas and wet unconsolidated sediments (Lagorio and Leal 2005). Nevertheless, many of them, particularly those located over lava flows, must be hydraulic breccias similar to those described in Brazil by Hartmann et al. (2010), as a consequence of the significant hydrothermal event of regional scale associated to this great volcanism.

Fig. 3.2 **a** Subhorizontal volcanic flows exposed at a road cut of Route 12. **b** Basaltic flow over a breccia level. **c** Detail of a peperitic breccia, also under the microscope; note the crenulations of the basaltic fragments as well as the remaining lamination of the matrix, both features characterizing pillowy peperites, as shown in Lagorio and Vizán (2011). **d** amethyst geode mineralization in Wanda locality

It should be noted that in a study performed in Brazil by Wildner et al. (2007) Serra Geral lavas were considered as a Group, and so they were assigned to different formations, essentially taking into account the geochemical lithotypes defined by Peate et al. (1992) as will be described below. This criterion has also been used in the mapping of some regions of Misiones by Remesal et al. (2011).

Although lava flows clearly dominate in the PMP, several sets of dykes (lava feeders) are well known in Brazil (e.g. Santos-Río de Janeiro, Ponta Grossa, Florianopolis) and Paraguay (Fig. 3.1a). In Misiones Province, dykes only have been reported at San Ignacio (Marengo and Palma 2005, Fig. 3.1b).

Samples presented here were collected in southern and northern sectors of the Misiones Province. The former includes rocks from the areas near the cities of Posadas, San Ignacio and Corpus, whereas the latter comprises samples from near the road between Bernardo de Irigoyen and Eldorado towns as well as a sample from Wanda locality (Figs. 3.1b and 3.2d). Sampling was guided by the field information provided by Ardolino and Mendía (1989), and results were given in Mena et al. (2006) and Lagorio and Vizán (2011). These data were analysed together with those obtained from Misiones in studies of the whole Paraná Magmatic Province, like those provided by Bellieni et al. (1986) and Piccirillo and Melfi (1988).

It should be noted that the paleomagnetic study of samples obtained from near the road between Bernardo de Irigoyen and Eldorado points out that the sequence has registered at least three polarity chrons (Mena et al. 2006). This is in agreement with results from the whole PMP (e.g. Ernesto and Pacca 1988), revealing that several polarity intervals are registered in the lava sequences. It is important to note that at the time of Paraná volcanism changes in polarity were highly frequent (e.g. Mena et al. 2006), which is in accordance with results obtained from Early Cretaceous alkaline lavas of Sierra Chica of Córdoba, as shown in Chap. 2.

3.1.2 Age of the Paraná Magmatic Province (PMP)

$^{40}Ar/^{39}Ar$ dating performed by Thiede and Vasconcelos (2010) points out that volcanism of the PMP began at 134.7 ± 1 Ma and it lasted less than 1.2 My. This value coincides with numerous previous data that yielded ages between 133 and 130 Ma (e.g. Renne et al. 1992, 1996a, 1996b; Marzoli et al. 1999; Ernesto et al. 1999), if recalculations according to Kuiper et al. (2008) are applied and therefore the former published ages become a bit older. This contrasts with the time span for the whole magmatic event, of about 11 million years, determined by other authors (Turner et al. 1994; Stewart et al. 1996; Turner et al. 1999a, b) on the basis of other $^{40}Ar/^{39}Ar$ dating. Recently, U-Pb ages obtained in interbedded acidic rocks range from 137.3 ± 1.8 to 134.3 ± 0.8 Ma (Wildner et al. 2006; Janasi et al. 2011), and a Re–Os isochron presented by Rocha-Júnior et al. (2012) yielded an age of 131.6 ± 2.3 Ma. Then, the age of the PMP is still a matter of debate, while there is a consensus that the volcanic event occurred during the Early Cretaceous.

As previously commented, in relation to the volcanism of the PMP, there are several alkaline localities marginally situated (Fig. 3.1c). According to radiometric dating provided by Gibson et al. (2006) and Comin-Chiaramonti et al. (2007), the Early Cretaceous potassic volcanism peripheral to the PMP is older (e.g. Amambay in north-eastern Paraguay), as well as contemporary (e.g. Jacupiranga, Ponta Grossa and Anitápolis from south-eastern Brazil; Mariscala in Uruguay) and also posthumous (Sapucai-Villarrica zone from eastern Paraguay) in relation to the large tholeiitic event. In addition, volcanism of Sierra Chica of Córdoba (Argentina) is slightly younger than PMP, according to the new age obtained (129.6 ± 1Ma, Chap. 2). Subsequently, ultrapotassic volcanism occurred in Late Cretaceous/Palaeogene times in northern Paraná basin (Gibson et al. 1995; Carlson et al. 1996, 2007).

3.1.3   Classification and Petrography

It is worth mentioning that a detailed petrographic and mineralogical study was presented sereral decades ago by Teruggi (1955).

The sampled rocks here presented are classified as basaltic andesites (mostly) and basalts, according to TAS diagram (Le Bas et al. 1986; Fig. 3.3a), as well as

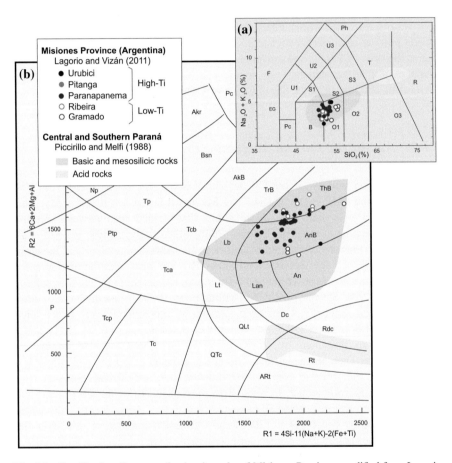

Fig. 3.3 Classification diagrams of volcanic rocks of Misiones Province, modified from Lagorio and Vizán (2011). **a** Alkali versus silica diagram (TAS, Le Bas et al. 1986); *B* basalt, *S1* trachybasalt, *S2* basaltic trachyandesite, *S3* trachyandesite, *T* trachyte, *F* foidite, *U1* basanite/tephrite, *U2* phonotephrite, *U3* tephriphonolite, *Ph* phonolite, *O1* basaltic andesite. **b** R1-R2 diagram (from De la Roche et al. 1980; Bellieni et al. 1981); $R1 = 4Si - 11 (Na + K) - 2$ (Fe + Ti), R2 = 6Ca + 2 Mg + Al; *Pc* picrite, *Akr* ankaratrite, *Bsn* basanite, *AkB* alkali basalt, *TrB* transitional basalt, *ThB* tholeiitic basalt, *Np* nephelinite, *Tp* tephrite, *Tcb* trachybasalt, *Lb* latibasalt, *AnB* andesi-basalt, *Ptp* phonotephrite, *Tca* trachyandesite, *Lt* latite, *P* phonolite, *Tph* trachyphonolite, *T* trachyte, *QTc* quartz-trachyte, *Lan* latiandesite, *An* andesite, *QLc* quartz-latite, *D* dacite, *Rd* rhyodacite, *R* rhyolite, *AR* alkaline rhyolite. Lithotypes of high- and low-Ti according to Peate et al. (1992)

andesi-basalts and tholeiitic basalts in R1-R2 (De la Roche et al. 1980; Bellieni et al. 1981; Fig. 3.3b). They present normative hypersthene and quartz, in accordance with their subalkaline tholeiitic chemistry.

The samples show textures ranging from aphyric (Fig. 3.4a) to microporphyritic (Fig. 3.4b) and porphyritic. The latter ones generally show a low proportion of phenocrysts and/or microphenocrysts (up to 5–8 %). Aphyric varieties are intergranular to intersertal, like the groundmass of microporphyritic and porphyritic rocks (Figs. 3.4a, b). The phenocrysts and/or microphenocrysts are labradoritic plagioclase (An_{50-65}), augite, opaque minerals (titanomagnetite and ilmenite), pigeonite, augite and olivine (frequently altered to iddingsite and/or bowlingite). This mineral assemblage also characterizes aphyric types (Fig. 3.4a, c). Pigeonite crystals are typical of tholeiitic lavas (Fig. 3.4c, d) in coexistence with augite. Titanomagnetite microphenocrysts often display ophitic texture (Fig. 3.4e), consistent with its late crystallization from subalkaline magmas.

The mesostasis consists of the aforementioned minerals, in addition to interstitial alkali feldspar and quartz sometimes forming micrographic intergrowths (Fig. 3.4f). Apatite is a common accessory phase. The glass frequently appears in variable proportions, fresh as well as replaced by mafic phyllosilicates as smectite–chlorite and/or celadonite. Clinopyroxenes, particularly pigeonite, may present a variable degree of uralitization. Celadonite is also common forming amygdales in lava flows (Fig. 3.4b), frequently along with quartz, silica-rich zeolites and carbonate. A peculiar assemblage composed of apophyllite, chabazite, chalcedony, quartz, stilbite, heulandite, celadonite and gypsum also forming amygdales has been described in the Freyer quarry, near Eldorado locality, by Latorre and Vattuone (1985). On the other hand, amethyst geode mineralization is conspicuously developed in the famous Wanda locality (Fig. 3.2d; Avila et al. 2008). This type of mineralization is well known in various localities of Brazil and Uruguay, and has an epigenetic origin within the significant hydrothermal event associated to Paraná volcanism, as pointed out by diverse authors (e.g. Duarte et al. 2009, 2014; Hartmann et al. 2010).

3.1.3.1 Comparison with Other Studies from Corrientes Province

Analyses carried out on volcanic rocks of Corrientes Province by Herrmann et al. (2011) indicate that samples correspond to basaltic andesites, basalts and andesites, according to TAS classification. Petrographic features are similar to those described for the Misiones Province and the whole PMP. It should be noticed that native copper was reported and studied in lavas from this province, as well as in other areas of the PMP. In Misiones Province copper was also described although no detailed study has been performed yet. Copper is related to primary silicates as well as to the alteration assemblage composed of smectitic and celadonitic phases, typical of these tholeiitic lavas. Also electrum occur in these rocks; besides, pyrite, chalcopyrite, digenite, cuprite and limonites have been reported (Herrmann et al. 2013). Furthermore, clinoptilolite and heulandite conforming microamygdales have been

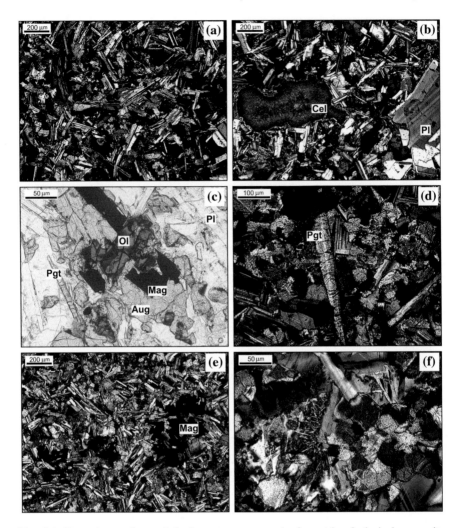

Fig. 3.4 Photomicrographs. **a** Aphyric texture composed of crystals of plagioclase, augite, pigeonite and opaque minerals in basaltic andesite of Paranapanema variety, surroundings of Posadas city. Crossed nicols, 4X. **b** Microporphyritic texture with plagioclase (*Pl*) microphenocrysts in an intergranular groundmass composed of microlites of plagioclase, augite, pigeonite and opaque minerals from a sample also bearing microamygdales filled with celadonite (Cel; *towards the left*) in a lava of Paranapanema variety from Profundidad area, near Posadas city, crossed nicols, 4X. **c** Crystals of plagioclase (*Pl*), augite (*Aug*), pigeonite (*Pgt*) and olivine (*Ol*) altered to bowlingite and iddingsite along with magnetite (Mag) in an aphyric pattern; basaltic andesite of Paranapanema variety from near Posadas city, parallel nicols, 20X. **d** Pigeonite (*Pgt*) crystal conforming the intergranular groundmass of a basaltic andesite of Paranapanema variety, surroundings of Posadas city, crossed nicols, 10X. **e** Magnetite (*Mag*) crystals enclosing microlites of plagioclase and clinopyroxenes in an ophitic pattern; basaltic andesite of Paranapanema variety from the surroundings of Posadas city, crossed nicols, 4X. **f** Micrographic intergrowths interstitially located in an aphyric basaltic andesite of Paranapanema variety from near Posadas city, crossed nicols, 20X

also described by the latter authors. Both primary and epigenetic copper mineralizations were therefore defined by Herrmann et al. (2011, 2013). Hydrothermal fluids must have removed copper carried by the tholeiitic lavas. The epigenetic genesis is in accordance with the characterization performed by Pinto et al. (2011) and Pinto and Hartmann (2014) in lavas of the PMP from Brazil. A significant hydrothermal event of regional scale must have taken place according to those Brazilian researchers, as above mentioned (e.g. Hartmann et al. 2010). Duarte et al. (2009, 2014) related the presence of alteration minerals as smectites, celadonite and zeolites along with amethyst geode mineralization and hydrothermal breccia formation, to fluids related to the Guaraní aquifer, at temperatures below 200 °C.

3.1.4 Geochemistry of Lavas from Misiones Province and Comparison with Magmas from the Whole Paraná Magmatic Province

Forty-five major and trace element chemical analyses are presented in Table 2.1. They were measured by RX fluorescence in the Department of Earth Sciences of the University of Trieste (Italy); nine of which were supplemented with rare earth measurements carried out by ICP-MS in Actlabs laboratory (Canada).

TiO$_2$ content has been proposed to discriminate the rocks from Paraná (Bellieni et al. 1984b), so the rocks can be of low Ti (<2 %, LTi) or high Ti (>2 %, HTi), afterwards generalized for all Mesozoic large continental basaltic flow provinces of Gondwana, as is well known.

Picirillo and Melfi (1988) pointed out from their own large database that low-Ti basalts predominate in the southern section of the Paraná region, low-Ti and high-Ti ones coexist in the central zone where intermediate lithotypes (2–3 % TiO$_2$) are common, while high-Ti basalts prevail in the northern area.

The samples analysed for the present study include both low- and high-Ti varieties, which is consistent since the Misiones Province belongs to the central and southern sections of the PMP.

Variation diagrams according to decreasing mg# show an increasing trend for SiO$_2$, FeOt, TiO$_2$, K$_2$O, Na$_2$O, P$_2$O$_5$, Zr, Nb, La, Ce, Nd, Rb and Ba and a diminishing tendency for CaO, Cr, Ni and Sr (Figs. 3.5 and 3.6). High-Ti rocks have higher contents of K$_2$O, P$_2$O$_5$, Zr, Nb, La, Ce, Nd, Sr and Ba, and lower Cr and Ni tenors than low-Ti rocks, for the same mg# (Figs. 3.5 and 3.6; Table 3.1).

Moreover, Peate et al. (1992) discriminated six varieties of basalts in PMP based on the Ti/Y ratio along with TiO$_2$ content, as well as Ti/Zr and Zr/Y ratios and Sr content. These varieties are called as follows: Urubici, Pitanga, Paranapanema, Ribeira, Gramado and Esmeralda. The first three varieties are considered of high Ti (TiO$_2$ values starting from 1.7 %), and the other three varieties are considered of low Ti (TiO$_2$ values can reach up to 2.3 %).

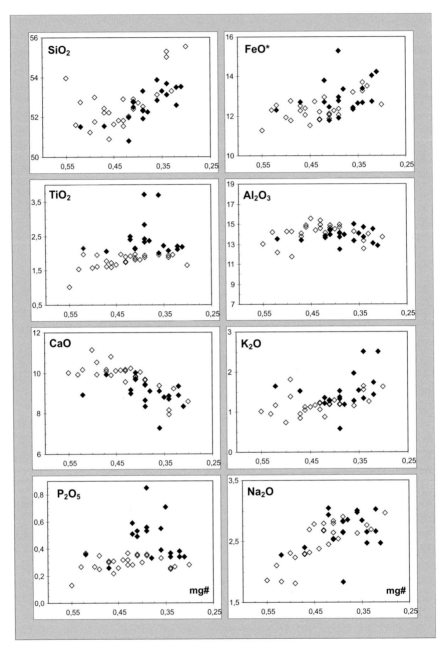

Fig. 3.5 Variation of major elements (wt%) with respect to magnesium number (mg#) for volcanic rocks of Misiones Province (MS). *Solid diamonds* high-Ti rocks, *open diamonds* low-Ti rocks. Based on Lagorio and Vizán (2011)

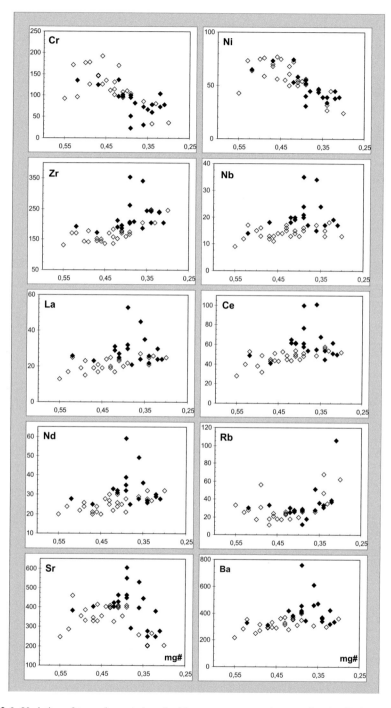

Fig. 3.6 Variation of trace elements (ppm) with respect to magnesium number (mg#). Symbols as in Fig. 3.5. Based on Lagorio and Vizán (2011)

Table 3.1 Major and trace element compositions of representative volcanic rocks of Misiones Province (Paraná Magmatic Province), enlarged from Lagorio and Vizán (2011)

Sample	M1	M2	M3	M4	M5	M6	M7	M8	M10	M10P	M11	M14	M15	M16	M17
Locality	PDS	PDS	PDS	PDS	PDS	PDS	PDS	PF	PF	PF	PF	AG	CPS	SI	SI
Rock	BA	B	B	BA	BA	BA	BA	BA	B	BA	BA	BA	BA	BA	BA
Type	PMA	PMA	PMA	PMA	PMA	PMA	PMA	PMA	PMA	PMA	PMA	PMA	PT	GR	GR
SiO_2	52.41	50.84	52.04	52.35	52.81	52.55	53.35	53.04	51.61	52.65	52.75	52.88	53.37	55.32	55.56
TiO_2	2.33	2.50	2.40	2.42	2.15	2.13	2.23	1.96	2.06	2.12	1.97	2.00	2.84	1.88	1.65
Al_2O_3	13.74	13.66	13.87	14.09	14.38	13.96	14.10	11.73	13.42	14.52	12.17	15.04	14.15	13.37	13.70
FeOt	12.95	13.81	12.72	12.77	11.77	12.47	12.64	12.79	12.73	12.76	12.55	12.45	11.9	13.26	12.59
MnO	0.21	0.16	0.17	0.21	0.17	0.19	0.20	0.22	0.19	0.21	0.17	0.20	0.17	0.21	0.21
MgO	4.31	5.18	4.80	4.23	4.37	4.63	3.57	6.36	5.87	3.21	6.83	3.65	3.98	3.52	2.82
CaO	9.45	9.19	9.00	9.43	10.04	9.72	8.81	9.93	9.93	9.37	10.17	9.13	8.66	7.97	8.60
Na_2O	2.65	2.93	3.05	2.64	2.54	2.53	2.84	1.81	2.40	3.03	1.84	2.97	2.83	2.65	2.96
K_2O	1.37	1.21	1.36	1.32	1.27	1.29	1.54	1.81	1.52	1.74	1.17	1.28	1.53	1.55	1.63
P_2O_5	0.56	0.51	0.59	0.53	0.49	0.53	0.71	0.35	0.26	0.38	0.37	0.39	0.56	0.26	0.28
Sum	100.00	100.00	100.00	100.00	100.00	99.99	100.00	100.00	100.00	100.00	100.00	100.00	100.00	99.99	100.00
mg#	0.40	0.43	0.43	0.40	0.43	0.43	0.36	0.50	0.48	0.34	0.52	0.34	0.40	0.35	0.31
L.O.I.	2.32	2.85	2.14	1.51	2.44	2.11	1.75	2.94	1.94	0.90	3.69	1.72	1.50	2.33	1.80
FeO	–	3.68	5.71	1.52	8.93	8.15	8.14	8.10	1.63	5.84	4.74	6.12	1.63	8.7	9.54
Fe_2O_3	–	11.24	7.78	12.49	3.14	4.79	4.99	5.20	12.32	7.68	8.67	7.02	11.40	5.06	3.38
Q (nor.)	4.00	0.13	1.70	4.24	4.77	4.25	5.41	4.46	0.84	2.22	5.09	3.64	5.81	8.36	7.34
Hy (nor.)	17.64	20.79	18.80	17.47	15.95	18.00	16.99	20.55	19.89	14.16	22.11	16.89	15.95	18.37	14.29
Cr	101	137	98	96	102	98	67	126	125	105	97	74	52	34	36
Ni	52	74	53	55	58	58	39	59	73	45	64	47	40	27	24
Rb	28	25	30	29	26	28	36	56	33		28	28	28	47	62
Ba	420	420	420	408	383	383	472	314	313	424	351	461	457	368	362

(continued)

Table 3.1 (continued)

Sample	M1	M2	M3	M4	M5	M6	M7	M8	M10	M10P	M11	M14	M15	M16	M17
Locality	PDS	PDS	PDS	PDS	PDS	PDS	PDS	PF	PF	PF	PF	AG	CPS	SI	SI
Rock	BA	B	B	BA	BA	BA	BA	BA	B	BA	BA	BA	BA	BA	BA
Type	PMA	PMA	PMA	PMA	PMA	PMA	PMA	PMA	PMA	PMA	PMA	PMA	PT	GR	GR
Sr	449	423	404	462	429		445	387	402	382	460	399	552	205	205
Nb	21	18	20	20	19	20	24	16	18	19	17	15	24	13	13
Zr	207	190	211	203	196	187	243	178	172	240	170	187	261	248	245
Y	32	28	30	30	29	29	35	27	22	30	25	31	33	40	41
La	32	29	31	30	25	27	31.8	23	23	30	25	24	32	25	25
Ce	58	62	65	61	61	62	68.7	49	41	62	53	55	77	58	52
Pr							8.24								
Nd	32	33	33	35	32	31	37.2	26	25	29	28	28	39	32	32
Sm							8.63								
Eu							2.62								
Gd							8.07								
Tb							1.38								
Dy							7.58								
Ho							1.50								
Er							4.13								
Tm							0.59								
Yb							3.76								
Lu							0.54								
Hf							5.6								
Ta							1.32								
U							0.71								
Th							3.58								

(continued)

Table 3.1 (continued)

Sample	M1	M2	M3	M4	M5	M6	M7	M8	M10	M10P	M11	M14	M15	M16	M17
Locality	PDS	PDS	PDS	PDS	PDS	PDS	PDS	PF	PF	PF	PF	AG	CPS	SI	SI
Rock	BA	B	B	BA	BA	BA	BA	BA	B	BA	BA	BA	BA	BA	BA
Type	PMA	PMA	PMA	PMA	PMA	PMA	PMA	PMA	PMA	PMA	PMA	PMA	PT	GR	GR
Pb							6								
Cu							178								
Zn							116								
Co							37								
V							369								
(La/Nb)$_{PM}$	1.58	1.67	1.6	1.56	1.37	1.40	1.38	1.49	1.33	1.64	1.53	1.66	1.38	2.00	2.00
(La/Yb)cn							5.70								
Ti/Y	436	535	480	484	444	441	382	436	562	424	508	387	516	282	241
Ti/Zr	67	79	68	72	66	68	55	66	72	53	75	64	65	45	40
Zr/Y	6.5	6.8	7.0	6.8	6.8	6.4	6.9	6.6	7.8	8.0	6.8	6.0	7.9	6.2	6.2
Sample	M18	M22	M31	M39	M40	M41	M42	M43	M44	M45	M46	M50	M62	M63	M64
Locality	SI	SI	SI	PDS	PDS	PDS	PDS	PDS	PDS	PDS	PDS	PDS	BI/ED	BI/ED	BI/ED
Rock	BA	BA	BA	BA	BA	B	BA	BA	BA	BA	BA	B	BA	BA	BA
Type	GR	UR	GR	PMA	PMA	PMA	PMA	RB	PMA	PMA	PMA	PMA	PMA	PMA	PMA
SiO$_2$	55.05	53.91	54.00	52.45	52.96	51.87	52.96	52.23	52.82	52.57	52.58	51.57	53.72	53.13	53.53
TiO$_2$	1.94	3.70	1.02	1.97	1.81	1.73	1.90	1.60	1.87	1.89	1.93	2.15	2.08	1.96	2.21
Al$_2$O$_3$	12.53	13.34	13.05	14.64	14.47	14.95	14.59	14.85	14.89	14.96	14.68	13.52	13.69	14.26	13.11
FeOt	13.72	12.34	11.27	12.12	12.07	12.22	11.8	12.06	11.87	12.07	12.36	12.30	13.41	13.21	14.06
MnO	0.21	0.16	0.20	0.20	0.18	0.19	0.18	0.19	0.19	0.19	0.19	0.16	0.21	0.21	0.21
MgO	3.74	3.70	7.43	4.4	4.44	4.86	4.71	5.02	4.30	4.14	4.22	7.06	3.65	3.72	35.6
CaO	8.16	7.30	10.02	9.83	9.86	10.13	9.58	9.91	9.71	9.71	9.67	8.94	8.88	9.38	8.89

(continued)

Table 3.1 (continued)

Sample	M18	M22	M31	M39	M40	M41	M42	M43	M44	M45	M46	M50	M62	M63	M64
Locality	SI	SI	SI	PDS	PDS	PDS	PDS	PDS	PDS	PDS	PDS	PDS	BI/ED	BI/ED	BI/ED
Rock	BA	BA	BA	BA	BA	B	BA	BA	BA	BA	BA	B	BA	BA	BA
Type	GR	UR	GR	PMA	PMA	PMA	PMA	RB	PMA	PMA	PMA	PMA	PMA	PMA	PMA
Na_2O	2.76	3.01	1.86	2.83	2.64	2.67	2.68	2.69	2.79	2.90	2.82	2.28	2.65	2.63	2.66
K_2O	1.64	1.97	1.02	1.19	1.20	1.06	1.23	1.13	1.21	1.20	1.19	1.65	1.34	1.15	1.43
P_2O_5	0.25	0.55	0.13	0.36	0.36	0.32	0.37	0.31	0.35	0.36	0.35	0.36	0.37	0.33	0.34
Sum	99.99	99.99	100.00	100.00	100.00	100.00	100.00	100.00	99.99	99.99	100.00	100.00	100.00	100.00	100.00
mg#	0.36	0.38	0.57	0.42	0.43	0.45	0.45	0.46	0.42	0.41	0.41	0.36	0.36	0.36	0.34
L.O.I.	1.52	2.13	3.54	1.49	2.42	1.25	1.97	2.00	1.16	1.77	1.56	1.11	1.11	1.11	0.92
FeO	6.77	3.81	3.10	9.52	10.47	3.07	8.37	7.64	1.47	8.39	9.35	1.47	1.47	1.47	4.40
Fe_2O_3	7.71	9.47	9.07	2.88	1.77	10.16	3.80	4.90	11.54	4.08	3.33	13.26	13.26	13.26	10.72
Q (nor.)	6.95	6.77	5.28	2.71	4.06	1.85	3.85	1.92	3.46	2.77	3.11	6.27	6.27	6.27	5.83
Hy (nor.)	17.78	15.46	24.18	16.81	17.15	18.61	17.96	19.19	16.89	16.63	17.10	17.62	17.62	17.62	17.19
Cr	33	31	93	102	108	132	102	126	96	104	105	61	61	86	74
Ni	32	44	43	55	53	61	50	56	50	56	51	34	34	46	38
Rb	68	51	33	26	27	23	24	24	27	26	27	31	31	26	37
Ba	363	613	219	357	371	314	371	334	354	396	348	340	340	306	318
Sr	203	532	247	404	410	410	398	414	397	410	401	251	251	258	251
Nb	15	34	9	15	14	15	17	13	14	16	15	17	17	17	19
Zr	246	340	131	170	177	137	182	150	176	173	166	242	242	204	239
Y	41	37	24	26	27	24	29	24	28	26	25	37	37	30	38
La	21	51.3	14.0	24	23	20	25	21	25	22	21.3	22	22	27	22.6
Ce	56	94.0	28.6	50	54	45	49	51	48	51	45.6	54	54	49	46.7
Pr		13.22	3.56								5.48				5.93
Nd	28	58.5	16.0	24	26	27	30	24	30	28	25.0	26	26	29	27.4

(continued)

Table 3.1 (continued)

Sample	M18	M22	M31	M39	M40	M41	M42	M43	M44	M45	M46	M50	M62	M63	M64
Locality	SI	SI	SI	PDS	PDS	PDS	PDS	PDS	PDS	PDS	PDS	PDS	BI/ED	BI/ED	BI/ED
Rock	BA	BA	BA	BA	BA	B	BA	BA	BA	BA	BA	B	BA	BA	BA
Type	GR	UR	GR	PMA	PMA	PMA	PMA	RB	PMA	PMA	PMA	PMA	PMA	PMA	PMA
Sm		12.6	3.89								5.78				6.51
Eu		3.81	1.19								1.86				2.03
Gd		10.8	4.04								5.75				6.90
Tb		1.72	0.71								0.98				1.24
Dy		8.94	4.19								5.46				7.48
Ho		1.63	0.84								1.12				1.49
Er		4.51	2.36								3.07				4.23
Tm		0.58	0.34								0.42				0.61
Yb		3.61	2.18								2.87				4.02
Lu		0.50	0.33								0.41				0.59
Hf		8.0	2.9								4.2				5.2
Ta		1.90	0.45								0.88				0.96
U		1.29	0.85								0.50				0.49
Th		5.46	3.61								2.59				2.78
Pb		49	8								5				17
Cu		176	111								258				257
Zn		146	66								111				123
Co		40	44								48				43
V		423	310								465				429
(La/Nb)PM	1.45	1.57	1.61	1.66	1.70	1.38	1.53	1.68	1.85	1.43	1.47	1.34	1.34	1.65	1.23
(La/Yb)cn		9.58	4.33								5.00				3.79

(continued)

Table 3.1 (continued)

Sample	M18	M22	M31	M39	M40	M41	M42	M43	M44	M45	M46	M50	M62	M63	M64
Locality	SI	SI	SI	PDS	PDS	PDS	PDS	PDS	PDS	PDS	PDS	PDS	BI/ED	BI/ED	BI/ED
Rock	BA	BA	BA	BA	BA	B	BA	BA	BA	BA	BA	B	BA	BA	BA
Type	GR	UR	GR	PMA	PMA	PMA	PMA	RB	PMA	PMA	PMA	PMA	PMA	PMA	PMA
Ti/Y	284	600	255	455	402	432	393	400	401	436	463	348	337	392	349
Ti/Zr	47	65	47	70	61	76	63	64	64	65	70	53	52	58	55
Zr/Y	6.0	9.2	5.5	6.5	6.6	5.7	6.3	6.3	6.3	6.7	6.6	6.5	6.5	6.8	6.5
Sample	M67	M68	M69	M71	M72	M74	M76	M77	M79	M80	M83	M84	M85	M87	M90
Locality	BI/ED	BI/ED	BI/ED	BI/ED	BI/ED	BI/ED	BI/ED	BI/ED	BI/ED	BI/ED	BI/ED	BI/ED	BI/ED	BI/ED	Wanda
Rock	BA	B	B	BA	BA	B	BA	B	B	B	BA	BA	BA	B	B
Type	PMA	PMA	RB	PMA	PMA	RB	PMA	RB	PMA	PMA	PMA	RB	PMA	RB	UR
SiO_2	52.29	50.95	51.27	53.57	53.18	51.61	52.76	51.80	51.84	51.60	52.28	52.48	52.07	51.68	51.91
TiO_2	2.36	1.72	1.58	2.18	2.08	1.55	1.81	1.61	1.97	1.75	1.77	1.62	1.91	1.66	3.72
Al_2O_3	13.94	14.70	14.25	12.88	14.70	14.23	14.51	14.25	14.39	15.35	14.07	13.76	14.14	15.55	12.45
FeOt	13.36	12.39	11.95	14.23	12.69	12.30	12.29	11.78	12.74	11.85	12.27	12.54	12.95	11.52	15.27
MnO	0.20	0.23	0.17	0.18	0.19	0.20	0.21	0.20	0.19	0.17	0.20	0.18	0.20	0.18	0.25
MgO	4.38	5.58	6.31	3.29	3.51	6.83	4.23	5.91	4.85	4.80	5.76	5.77	4.89	5.13	4.79
CaO	9.11	10.83	11.15	8.34	8.72	9.94	10.05	10.56	10.15	10.18	10.10	10.19	10.23	10.11	8.33
Na_2O	2.85	2.32	2.31	2.47	3.05	2.11	2.54	2.24	2.38	2.79	2.29	2.30	2.45	2.78	1.83
K_2O	1.18	1.05	0.74	2.51	1.51	0.96	1.29	1.39	1.17	1.22	0.95	0.85	0.88	1.12	0.59
P_2O_5	0.33	0.22	0.27	0.34	0.36	0.27	0.30	0.25	0.32	0.28	0.30	0.31	0.28	0.26	0.85
Sum	100.00	100.00	100.00	100.00	99.99	100.00	100.00	100.00	100.00	100.00	100.00	100.00	100.00	100.00	100.00
mg#	0.40	0.47	0.52	0.32	0.36	0.53	0.41	0.50	0.44	0.45	0.49	0.48	0.43	0.47	0.39
L.O.I.	1.64	2.01	2.77	1.50	1.76	4.41	2.69	2.50	4.71	1.76	2.70	3.18	1.91	2.88	6.75
FeO	7.28	6.39	5.23	1.59	6.53	1.32	6.11	6.93	0.90	2.36	0.87	5.92	6.89	1.48	4.84

(continued)

Table 3.1 (continued)

Sample	M67	M68	M69	M71	M72	M74	M76	M77	M79	M80	M83	M84	M85	M87	M90
Locality	BI/ED	BI/ED	BI/ED	BI/ED	BI/ED	BI/ED	BI/ED	BI/ED	BI/ED	BI/ED	BI/ED	BI/ED	BI/ED	BI/ED	Wanda
Rock	BA	B	B	BA	BA	B	BA	B	B	B	BA	BA	BA	B	B
Type	PMA	PMA	RB	PMA	PMA	RB	PMA	RB	PMA	PMA	PMA	RB	PMA	RB	UR
Fe_2O_3	6.74	6.66	7.46	14.03	6.83	12.19	6.86	5.38	13.14	10.53	12.66	7.34	6.72	11.14	11.57
Q (nor.)	3.09	0.67	1.19	4.34	3.11	2.05	3.88	1.28	3.00	0.49	3.60	3.88	3.57	0.52	11.07
Hy (nor.)	18.52	19.49	20.23	16.63	16.47	24.72	16.53	19.46	18.68	17.67	20.84	21.05	18.87	18.67	24.26
Cr	83	192	177	79	64	171	111	178	112	115	145	147	170	135	23
Ni	45	77	75	39	38	73	54	76	55	68	70	68	72	75	31
Rb	18	21	17	106	33	25	20	31	17	26	18	11	18	18	12
Ba	345	295	252	337	366	285	310	269	363	327	296	293	279	290	758
Sr	292	331	354	280	297	286	354	332	404	401	330	347	327	355	604
Nb	17	11	15	17	19	12	13	13	14	16	13	12	13	14	35
Zr	208	144	143	204	213	170	159	142	171	157	153	145	153	137	354
Y	32	24	20	29	35	28	27	23	24	24	28	25	26	24	39
La	21	17	19	24	26	17	20	14.6	24	19	16.4	19	16.5	19	36.6
Ce	54	44	38	50	45	40	45	30.6	53	49	34.0	43	35.2	43	81.0
Pr								3.89			4.26		4.29		9.82
Nd	25	21	22	28	27	24	25	17.6	28	25	20.1	21	20.1	20	45.1
Sm								4.18			4.78		4.81		10.3
Eu								1.44			1.60		1.63		3.19
Gd								4.21			4.95		5.04		9.47
Tb								0.76			0.85		0.87		1.56
Dy								4.35			5.14		5.14		8.17
Ho								0.87			1.02		1.04		1.60
Er								2.54			2.92		2.93		4.31

(continued)

Table 3.1 (continued)

Sample	M67	M68	M69	M71	M72	M74	M76	M77	M79	M80	M83	M84	M85	M87	M90
Locality	BI/ED	BI/ED	BI/ED	BI/ED	BI/ED	BI/ED	BI/ED	BI/ED	BI/ED	BI/ED	BI/ED	BI/ED	BI/ED	BI/ED	Wanda
Rock	BA	B	B	BA	BA	B	BA	B	B	B	BA	BA	BA	B	B
Type	PMA	PMA	RB	PMA	PMA	RB	PMA	RB	PMA	PMA	PMA	RB	PMA	RB	UR
Tm								0.34			0.43		0.42		0.55
Yb								2.34			2.74		2.83		3.67
Lu								0.34			0.41		0.41		0.52
Hf								3.1			3.5		3.6		7.5
Ta								0.62			0.66		0.64		1.79
U								0.30			0.38		0.44		0.78
Th								1.76			2.27		2.23		3.93
Pb								-5			17		6		8
Cu								192			115		227		211
Zn								65			101		101		147
Co								42			44		46		37
V								368			384		404		376
(La/Nb)$_{PM}$	1.28	1.60	1.31	1.47	1.42	1.47	1.60	1.17	1.78	1.23	1.31	1.64	1.32	1.41	1.08
(La/Yb)cn								4.21			4.04		3.93		6.72
Ti/Y	442	430	474	451	357	332	402	420	492	438	379	389	441	415	572
Ti/Zr	68	72	66	64	59	55	68	68	69	67	69	67	75	73	63
Zr/Y	6.5	6.0	7.1	7.0	6.1	6.1	5.9	6.2	7.1	6.5	5.5	5.8	5.9	5.7	9.1

Major elements recalculated to 100 % on a volatile free basis. mg-no = $Mg/(Mg + Fe^{2+})$ and CIPW-normative compositions assuming $Fe_2O_3/FeO = 0.15$. Q, Hy = CIPW-normative quartz and hypersthene, respectively. PM = normalized to Primordial Mantle of Sun and McDonough (1989). cn = chondrite normalized according to Boynton (1984). Trace element contents in normal and italic styles indicate measurement by XRF and ICP-MS techniques, respectively. *PDS* Posadas, *PF* Profundidad, *AG* Arroyo Guarupa, *CPS* Corpus, *SI* San Ignacio, *BI/ED* Bernardo de Irigoyen/Eldorado transect. *BA* basaltic andesite, *B* basalt, PMA Paranapanema, *PT* Pitanga, GR Gramado, *RB* Ribeira, *UR* Urubici

The samples of Misiones Province that were herein analysed were classified as Paranapanema (most of them), Ribeira, Gramado, Pitanga and Urubici (the last two varieties are very scarce, see Table 3.1).

Rocks from the Eldorado–Bernardo de Irigoyen section (northern part of the Misiones Province, Fig. 3.1b) are typified as Paranapanema and Ribeira, and those around Posadas (southern area) are characterized as Paranapanema, Gramado, Urubici and Pitanga (Table 3.1). A sample from the Wanda locality (north of Eldorado, Fig. 3.1b) belonging to a lava flow bearing amethyst geodes has also been included, despite its high loss of ignition. The last feature seems to characterize volcanic levels displaying such cavity fillings (Hartmann pers. comm. 2010), and according to TiO_2 content, Ti/Y, Ti/Zr and Zr/Y ratios the sample can be classified as of the Urubici type. Although the Esmeralda lithotype has not been recognized in the present sampling, it has been found in few areas of the central part of Misiones, where Paranapanema type prevails (Remesal et al. 2011).

Peate et al. (1992) pointed out that low-Ti Gramado and Esmeralda as well as high-Ti Urubici rocks are characteristic of southern Paraná, whereas high-Ti Pitanga and Paranapanema as well as low-Ti Ribeira rocks are typical of northern Paraná region, while all varieties coexist in the central sector.

In multi-element diagrams normalized to primordial mantle (Sun and McDonough 1989), negative anomalies for Nb-Ta, Sr, P and Ti, typical of continental basalts, can be seen (Fig. 3.7a, b). High-Ti rocks are characterized by higher trace element contents than low-Ti ones (Fig. 3.7a, b), particularly the Urubici sample (Fig. 3.7b, c). Gramado samples display pronounced negative anomalies for Nb, Ta, Sr, P and Ti (Fig. 3.7b, d) as pointed out in the literature. Although this lithotype is characterized by the great LILE enrichment over HFSE and LREE, the sample plotted here shows very high Rb, Th, U and K tenors (Fig. 3.7b, d). The latter clearly becomes evident when comparing with low-Ti samples from southern Paraná that display $^{87}Sr/^{86}Sr$ values <0.7060 (Fig. 3.7d) presented by Ernesto et al. (2002).

REE diagrams normalized to chondrite (e.g. Boynton 1984) display flat patterns for the samples (Fig. 3.8), revealing low LREE/HREE fractionation (La/Ybcn between 3.79 and 9.58, Table 3.1). However, the latter tend to be higher for high-Ti rocks, particularly for Urubici variety (Fig. 3.8; Table 3.1). It should be noted that this is essentially a consequence of the higher content of La rather than a variation in Yb tenors.

3.1.4.1 Comparison with Other Geochemical Studies from Corrientes Province

Analyses carried out on these lavas in Corrientes Province by Herrmann et al. (2011) indicate that samples mainly belong to the Gramado and Esmeralda varieties, the former being the prevailing lithotype; anyway, also a few samples of Paranapanema and Pitanga occur. The observed varieties are in accordance with those reported by other authors from the whole Paraná basin, as low-Ti magmas

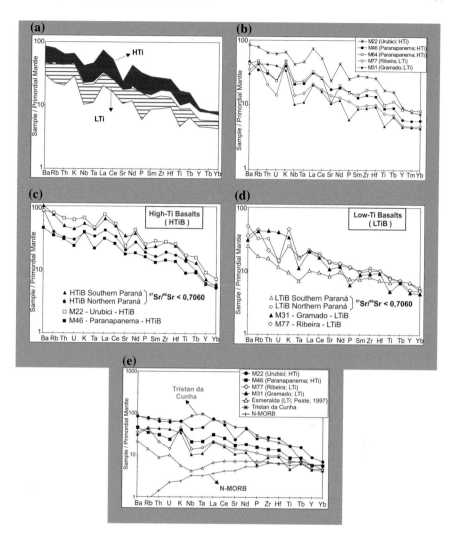

Fig. 3.7 Primordial mantle (Sun and McDonough 1989) normalized multi-element plots for volcanic rocks of Misiones Province (MS), as shown in Lagorio and Vizán (2011). **a** Discrimination according to Ti content, **b** diagram considering the lithotypes defined by Peate et al. (1992), **c** and **d** high- and low-Ti varieties from samples of MS in comparison with average compositions of high-Ti and low-Ti rocks with $^{87}Sr/^{86}Sr < 0.706$ from northern and southern regions of Paraná Magmatic Province (PMP) presented by Ernesto et al. (2002). **e** Comparison of samples of MS with those of Tristan da Cunha (le Roex et al. 1990) and N-MORB compositions, similarly as it was presented by Ernesto et al. (2002) for the whole PMP

prevail in southern PMP (Piccirillo and Melfi 1988; Peate et al. 1992), already mentioned above.

Herrmann et al. (2011) pointed out the high levels of Cu carried by the studied basalts in Corrientes Province, with an average of 170 ppm, ranging from 85 to

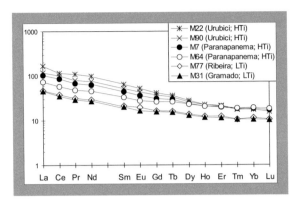

Fig. 3.8 Chondrite-normalized (Boynton 1984) REE patterns of selected volcanic rocks of MS, adapted from Lagorio and Vizán (2011)

385 ppm. It should be taken into account that contents of 100–200 ppm of Cu are considered to be enriched in that element by Cox (1986).

In a similar way, Cu tenors in basalts from the Misiones Province display an average of 192 ppm, varying from 111 to 258 ppm (presented in this study, Table 3.1). Cu contents of lavas of both Misiones and Corrientes Provinces display maximum values higher than 250 ppm, as in southern Brazil, western and central Uruguay, according to comparison with data presented by other authors (e.g. Peate et al. 1999; Turner et al. 1999a, b) as indicated by Herrmann et al. (2011).

3.1.5 Petrogenetic Aspects of Lavas of Misiones Province in the Context of Paraná Magmatic Province

3.1.5.1 Differentiation Processes

Samples from the province of Misiones (Table 3.1) have relatively low MgO content (7.43–2.82 %) and mg# values (0.54–0.29) along with low-Ni tenors (75–24 ppm), in accordance with PMP data from the literature (e.g. Piccirilo and Melfi 1988; Peate 1997), which clearly demonstrate that these lavas do not represent primary magmas.

Major and trace element variation diagrams depict evolutionary trends in agreement with fractional crystallization processes, for both high-Ti and low-Ti basalts (Figs. 3.5 and 3.6). Modelling was performed with MELTS software (Ghiorso and Sack 1995) considering low pressure (1 kbar) and almost anhydrous (0.5 % H_2O), as selected by Iacumin et al. (2003). The latter seem to be suitable as the prevailing phenocrystal assemblage is characterized by plagioclase and scarce olivine. As is well known, high H_2O must have inhibited plagioclase fractionation (e.g. Barberi et al. 1971). Low H_2O content also determines low oxygen fugacity that promotes the delay in magnetite crystallization, consistent with FeO and TiO_2

(a) **(b)**

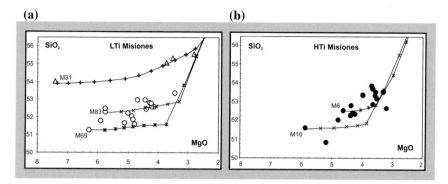

Fig. 3.9 MgO versus SiO₂ trends obtained with MELTS software (Ghiorso and Sack 1995) for selected samples of LTi and HTi (**a, b**, respectively) of the province of Misiones, applying the conditions selected by Iacumin et al. (2003), shown in Lagorio and Vizán (2011). *Open circles* LTi rocks (<2 % TiO₂, Ribeira and some samples of Paranapanema varieties), *open triangles* LTi Gramado-type rocks, *filled circles* HTi samples

increasing trends in variation diagrams (Fig. 3.5), so that QFM buffer was considered. Liquid lines of descent were obtained from compositions of LTi basalts (M69 and M83, Ribeira variety; M31, Gramado type) and HTi basalts (M6 and M10, Paranapanema variety), illustrated for MgO versus SiO₂ in Fig. 3.9. MELTS fractional crystallization trends are consistent with the general evolution of magmas of low and high titanium, respectively (Fig. 3.9a, b). Mineral fractionation sequence is composed of clinopyroxene, plagioclase and magnetite, in accordance with the petrography and the geochemical trends. In Fig. 3.9a, it is also remarkable that samples corresponding to Gramado type are distinguished because of the high SiO₂ contents, but it should be noted that they also support fractional crystallization by the trend obtained with MELTS. Nevertheless, high tenors of Rb, U, Th (Table 3.1; Fig. 3.7b, d) seem to point out crustal participation, consistent with the high $^{87}Sr/^{86}Sr$ typical of this variety.

3.1.5.2 Mantle Source

As it has been previously mentioned, a considerable difference in trace element contents (e.g. Ti, P, Sr, La, Ce, Nd, Zr, Nb, Ba) distinguishes high- and low-Ti basalts mainly reflecting distinct features in the mantle sources, as differentiation processes cannot account for such differences. Also Sr-Nd isotopic data from literature (Fig. 3.10) and Pb isotopes (e.g. Marques et al. 1999) point out that the source region of both types of tholeiites seems to have been different. Surprisingly, eleven isotopic data of Sr obtained by Rocha-Júnior et al. (2012) do not show any difference between high- and low-Ti basalts.

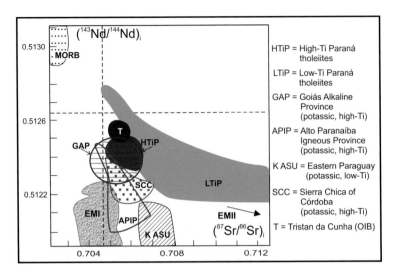

Fig. 3.10 Initial $^{87}Sr/^{86}Sr$ versus $^{143}Nd/^{144}Nd$ of high- and low-Ti tholeiites from the Paraná Magmatic Province (PMP; data: Piccirillo and Melfi 1988; Peate and Hawkesworth 1996; Peate 1997). Also data from potassic localities around Paraná basin are included (*Goiás Alkaline Province* Carlson et al. 1996, 2007; *Alto Paranaíba Igneous Province* Gibson et al. 1995, Carlson et al. 1996, 2007; *Eastern Paraguay* Comin-Chiaramonti et al. 1997, 2007; *Sierra Chica of Córdoba* Kay and Ramos 1996, Lucassen et al. 2002), as well as from Tristan da Cunha (le Roex et al. 1990). Data of MORB, EMI and EMII are taken from Zindler and Hart (1986) for comparison

Chondrite-normalized (cn) Yb concentrations in the studied rocks of MS are higher than 10 (10–20, Fig. 3.8, Table 3.1), and $(La/Yb)_{cn}$ varies between 3 and 10. It should be noted that high-Ti rocks tend to display higher $(La/Yb)_{cn}$ ratios (3.79–9.51) than low-Ti ones (3.93–5), as shown in Table 3.1. Ratios are more influenced by variations in La rather than in Yb content, consistent with diverse enrichments in the mantle source as well as variations in melting degrees.

For the characterization of simple melting process, mantle source of HTi was modelled, particularly for the most enriched type (Urubici lithotype). The composition of "primary" magmas of this variety was calculated from sample M22 (Table 3.2), and melting degree was obtained through XLFRAC starting from an enriched peridotitic major element composition (e.g. pyrolite of Ringwood). The obtained melting degree is 10.45 %, considering an anhydrous residual garnet-bearing peridotite (Table 3.2). As is well known, from a less enriched mantle compositions, melting degrees become lower but ≥5 %, in agreement with the tholeiitic nature of the lavas (e.g. Frey et al. 1978; Jaques and Green 1980; Takahashi and Kushiro 1983).

Trace element contents in the mantle source were calculated (Table 3.2), assuming batch melting (Hanson 1978). In a multi-element diagram (Fig. 3.11), it can be seen that the source of HTiB is enriched with respect to the primordial

Table 3.2 Calculated major and trace element content of primary magma (PM) from high-Ti Urubici-type (M22)

	PM M22		PM M22		P1	PM M22 res.		Source M22
SiO_2	51.47	Ba	383.37	OL^1	56.23	58.40	Ba	40.98
TiO_2	2.18	Rb	31.93	OPX^2	15.61	12.42	Rb	3.39
Al_2O_3	8.37	K	10,127.12	CPX^2	16.35	10.77	K	1064.18
FeOt	10.91	Nb	21.58	GT^1	11.81	7.97	Nb	2.71
MnO	0.09	La	31.95	$\Sigma\ res^2$	0.62	0.34	La	3.57
MgO	16.61	Ce	59.31	F (%)		**10.45**	Ce	6.93
CaO	7.17	Sr	335.64				Sr	37.90
Na_2O	1.75	Nd	37.35	P2	PM4 res.		Nd	4.66
K_2O	1.13	Sm	8.03	OL	54.49	58.48	Sm	1.20
P_2O_5	0.32	Zr	215.07	OPX	18.12	18.37	Zr	28.98
Sum	100.00	Ti	14,091.17	CPX	11.37	9.84	Ti	1806.63
		Y	24.06	GT	13.16	10.80	Y	6.85
mg#	0.73	Yb	2.34	PHL	2.86	2.50	Yb	0.86
				$\Sigma\ res^2$	0.40	0.30		
				F (%)		**4.99**		

Compositions were calculated to match at $100*Mg/(Mg + Fe^{2+})$ of 0.73, as performed by Iacumin et al. (2001). Primary magma of M22 = M22 + 30 % OL(89) + 20 % CPX. Weight percent of mineral assemblage (Stormer and Nicholls 1978) from peridotite composition of Ringwood (1966; P1) and mineral assemblage of residua after extraction of PM M22 (primary magma of M22). Incompatible element content (batch melting; Hanson 1978) of mantle source of PM M22 is also calculated. A primary-type magma of high-Ti alkaline volcanism of SCC (see Table 2.9) is also included for comparison

[1]McGregor (1974), [2]Ringwood (1966). OL olivine; OPX orthopyroxene; CPX clinopyroxene; GT garnet; PHL phlogopite; F melting degree; res. residua

[b]Figures in bold correspond to F (melting degree). Bold type was used to mark the difference between the values required to obtain tholeiitic and alkaline magmas, starting from similar mantle sources

mantle of Sun and McDonough (1989) and that enrichment can be up to 4–7 times respect to that mantle (e.g. for Ba, Rb, K, Nb, La, Ce and Nd).

It should be taken into account that HTi basalts of PMP present similar isotopic composition respect to the high-Ti alkaline rocks from peripheral localities to Paraná basin (Fig. 3.10), revealing analogies between both mantle sources. The comparison performed between sources of Urubici type of Misiones and alkaline basalts of SCC, presented here and previously attempted by Lagorio (2008), shows a more enriched character of the source of the alkaline basalt of SCC respect to LILE, Nb and Zr (Fig. 3.11a, b); in this case, the enrichment related to the primitive mantle of Sun and McDonough (1989) may be up to 10–20 times for Ba and K.

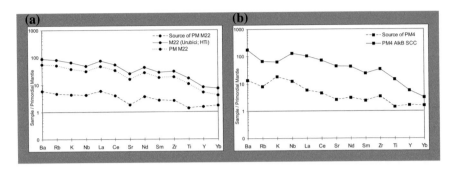

Fig. 3.11 a Multi-element plot of calculated (batch melting; Hanson 1978) mantle source, normalized to primitive mantle (PM, Sun and McDonough 1989), for the calculated primary magma of a HTi Urubici-type sample (M22), starting from an enriched peridotite (pyrolite; Ringwood 1966) with residual garnet in the source. The calculated primary magma is included. **b** Multi-elemental plot of calculated mantle source, normalized to primitive mantle (Sun and McDonough 1989), for a primary magma of a high-Ti alkali basalt (PM4) from SCC, starting from an enriched peridotite (pyrolite; Ringwood 1966) with residual garnet and phlogopite in the source, shown in Fig. 2.20 for comparison

3.1.6   Petrogenetic Aspects of Tholeïtes of Misiones Province (North-eastern Argentina) in the Context of PMP. Relationship with the Source of the Alkaline Volcanism of Córdoba Province (Central Argentina)

Samples from the Misiones Province presented in this study are of high- and low-Ti varieties, in accordance with the geographic location of this locality that belongs to the central and southern Parana Magmatic Province. Paranapanema constitutes the most abundant lithotype in the studied samples. Urubici variety, here recognized, represents the westernmost outcrops of this type in the region.

Rocks from the province of Misiones indicate that evolution of HTi and LTi magmas must have occurred from two different parental magmas through fractional crystallization at low pressures, with a low water content and low oxygen fugacity, involving crustal contamination only for the Gramado variety in agreement with the general trend recognized in the PMP.

Chondrite-normalized (cn) Yb concentrations in the studied rocks of Misiones are higher than 10, showing quite flat patterns in chondrite-normalized diagrams (Fig. 3.8; Table 3.1). Although these features seem to indicate that garnet was absent as a residual phase in the source, $(La/Yb)_{cn}$ higher than 5 presented by most of HTi basalts from Misiones must be more consistent with garnet retention in the mantle, according to what was determined by Iacumin et al. (2003) for the whole PMP.

It is also significant that La/Nb ratios normalized to primordial mantle (PM) are >1 (1–2, Table 3.1) determining the negative anomaly for Nb, also present for Ta, typical feature of the source of Paraná magmas. It should be noted that considering the whole Paraná samples, only a low proportion do not show Nb negative anomaly, corresponding to samples of LTi bearing $(La/Nb)_{PM} \leq 1$ and low (1–2) $(La/Yb)_{cn}$ with very flat REE patterns, that seem to be compatible with a garnet-free peridotite or a residual spinel-bearing source, as indicated by Iacumin et al. (2003). This type does not appear in samples from Misiones, since LTi basalts bear $(La/Yb)_{cn} < 5$ (3–5), but they display $(La/Nb) \geq 1$); on the other hand, most HTi basalts show high $(La/Yb)_{cn}$ and all have $(La/Nb)_{PM}$ ratios ≥ 1, so garnet as residual phase in the mantle source is supported.

Characterization of mantle source of the PMP from a tectonic point of view has been a matter of debate among the diverse research groups through the last decades. The so-called Tristan plume was strongly invoked as the cause of PEA LIP (e.g. White and McKenzie 1989; Courtillot et al. 2003). To test for this hypothesis, the chemical composition of the present-day Tristan da Cunha lavas (le Roex et al. 1990) was compared with that of the Paraná lavas by Ernesto et al. (2002). In the same way, in this contribution, data from Misiones Province belonging to the diverse varieties of Peate et al. (1992) were plotted in a normalized multi-element diagram (Fig. 3.7e). The essential difference with Tristan da Cunha pattern lies in the Nb-Ta anomaly, as pointed out by Ernesto et al. (2002); whereas for the rocks of Misiones Province and the majority of the samples from the Serra Geral Formation of the PMP that anomaly is negative, for Tristan it is clearly positive. It is noteworthy that Os isotopic data presented by Rocha-Júnior et al. (2012) also indicate no relation between Paraná basalts and present-day Tristan da Cunha lavas.

The comparison between Misiones data and N-MORB composition shows significant differences particularly from Rb to P (Fig. 3.7e), as was also pointed out by Ernesto et al. (2002) for the PMP. However, the Esmeralda variety (a representative sample taken from Peate 1997 was included) is characterized by a remarkable depletion in incompatible trace elements and therefore could represent mixing with an N-type MORB component, as has been suggested by Peate and Hawkesworth (1996) considering also the isotopic ratios of that variety.

It should be noted that geochemical characterization based on Ti and its distribution in a regional context (prevalence of high-Ti basalts in the north and low-Ti in the south of PMP) reveal differences in the mantle source, consistent with large-scale heterogeneities, typical of the subcontinental lithospheric mantle (SCLM), as strongly claimed by several researchers (e.g. Piccirillo and Melfi 1988; Peate 1997). Surprisingly, the recent Os isotopic data indicate that there is no evidence of ancient SCLM as a major source component in the genesis of the Paraná basalts, although the possibility cannot entirely be ruled out (Rocha-Júnior et al. 2012). They proposed, instead, that essentially the sublithospheric mantle has been involved in the genesis of PMP.

The distinctive Nb-Ta negative anomaly of PMP basalts is a typical feature of subduction-related magmas, as well as the high La/Nb and Ba/Nb ratios, as is well known. As PMP basalts were extruded in an extensional setting during Early Cretaceous times, related to the onset of Gondwana break-up, the subduction geochemical signature can only be interpreted as inherited from older subduction processes, whose related fluids must have metasomatized the mantle source in the geological past, as indicated by diverse authors. Furthermore, retention of Nb, Ta and HREE by the structure of rutile, ilmenite and garnet of eclogites from ancient slabs hosted in the mantle of PMP basalts can account for the Nb-Ta negative anomaly and high La/Nb and Ba/Nb ratios, as pointed out by Iacumin et al. (2003). According to these authors, also $(La/Yb)_{cn}$ ratios could be increased by garnet from eclogitic–granulitic subducted material, particularly in the HTi Paraná basalts. It should be also taken into account that, in the same way, La and Ba contents must have been enlarged by the presence of recycled sediments associated to a slab, which determined source enrichment, along with other LIL and LRE elements.

Besides, it is worth mentioning that isotopic data also allowed other research groups to identify mixing processes in the source of Paraná basalts related to ancient subduction processes (e.g. Transamazonian and Brasiliano events, Iacumin et al. 2003 or just the Brazilian event, Rocha-Júnior et al. 2012, 2013).

Precambrian subduction processes in the mantle source of MS and the whole PMP are also supported by Nd model ages which point out that metasomatism in the source of low-Ti basalts must have taken place between 2.4 and 0.7 Ga, whereas during 1.7–0.9 Ga in the source of high-Ti magmas (Comin-Chiaramonti and Gomes 2005; Gastal et al. 2005; Comin-Chiaramonti et al. 2007).

The subsurface of Misiones Province is in a zone between the Río de la Plata craton and the Paranapanema or Paraná block, with the Paraguay belt located between them (e.g. Bossi and Gaucher 2004; Rapela et al. 2011; Gaucher et al. 2011; Casquet et al. 2012) (Fig. 3.12b). Therefore, ancient subduction processes must have taken place, mainly during Late Proterozoic–Early Cambrian times. Juvenile Cambrian arc rocks (540–510 Ma) from the boundary of Paraguay and Brasilia belts are exposed in Brazil (Ferreira et al. 2008). It should be taken into account that also subduction from ancient terranes could also have affected the source of PMP basalts. For example, concerning the Misiones Province area, the Nico Pérez terrane (Archaean to Mesoproterozoic unit reworked during the Neoproterozoic) could extend up to northern Paraguay (Bossi and Gaucher 2004; Gaucher et al. 2011; Fig. 3.12b) and presents geological evidence of subduction from 3.1 Ga to at least 1.25 Ga (Gaucher pers. comm. 2013). On the other hand, other researchers consider, instead, the Río Apa block (e.g. Rapela et al. 2007, 2011; Casquet et al. 2012) as a unit that evolved during Mesoproterozoic times, and it is located in Paraguay, to the NW of Misiones (Fig. 3.12a). Regardless of those proposed models, it can be inferred that even the most ancient subduction processes could have affected the mantle source of PMP basalts.

As stated before, also lithospheric mantle source is interpreted for the alkaline volcanic rocks of SCC, a feature which also characterizes the other potassic localities around the Paraná basin.

As also mentioned, volcanism of the SCC is overlying the Pampean mobile belt, which is coeval with other Brasiliano orogens in Brazil (e.g. Paraguay and Araguaia fold belts) active in Neoproterozoic–Early Cambrian (e.g. Escayola et al. 2007), or during Early–Middle Cambrian times (e.g. Thover et al. 2010; Rapela et al. 2011; Gaucher et al. 2011), whereas APIP and Goiás alkaline localities are overlying the Brasilia belt. Subduction-related granitoids outcrop north of the Sierra Chica (in Sierra Norte of Córdoba), with ages of 537 ± 4 Ma and 530 ± 4 Ma (Iannizzotto

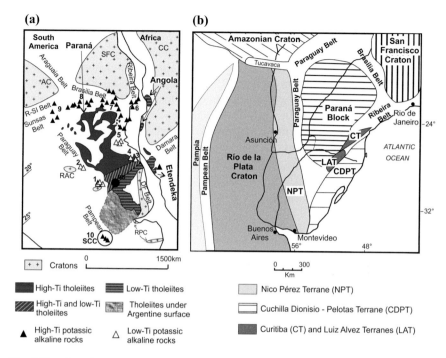

Fig. 3.12 a Location of the PMP including the alkaline peripheral localities, with discrimination according to Ti content, in the south-western part of Gondwana at about 130 Ma. Adapted from Piccirillo and Melfi (1988), Gibson et al. (1996), Marzoli et al. (1999), Lagorio (2008) and Lagorio and Vizán (2011). *MS* Misiones, *SCC* Sierra Chica of Córdoba, *RPC* Río de la Plata craton, *SFC* San Francisco craton, *AC* Amazonian craton, *CC* Congo craton, *RAC* Rio Apa craton. *1* eastern Paraguay, *2* Amambay, *3* Anitápolis, *4* Lages, *5* Ponta Grossa, *6* Serra do Mar, *7* Alto Paranaíba, *8* Goiás, *9* Poxoreu. **b** Major tectonostratigraphic units of SE-South America, adapted from Bossi and Gaucher (2004) and Gaucher et al. (2011)

et al. 2013), associated with Cambrian to Ordovician volcanic and subvolcanic bodies of calc-alkaline nature (O'Leary et al. 2014). Nevertheless, the volcanic rocks of SCC do not reveal chemical parameters that could be easily related to slab-derived mantle metasomatism, as neither do many ones from Brazil. Therefore, the ancient subduction must have not contaminated the source of the magmas of SCC; precisely, Nd model ages obtained by Lucassen et al. (2002) range from 1.22 to 0.96 Ga, suggesting a reactivation of the lithospheric mantle modified during Mesoproterozoic times, as pointed out previously (Lagorio 2008).

Subtle chemical heterogeneities were considered to be compatible with a lithospheric mantle source by Lagorio (2008). Metasomatic processes might have been produced, therefore, by small volumes of volatile-rich melts ascending from the asthenosphere, as reported in many of the alkaline localities around PMP (e.g. Gibson et al. 1995, 2006; Comin-Chiaramonti et al. 1997, 2007, 2013).

According to Os isotopic data, Rocha-Júnior et al. (2013) proposed that the sources of HTi alkaline basalts of northern Brazil, physically close to Paraná lavas, might have been spatially continuous, and therefore, both affected by metasomatism related to Neoproterozoic subduction processes. On the other hand, to the south-west of PMP, high-Ti SCC alkaline basalts are close to the inferred LTi basalts from the PMP that are lying under the surface (Fig. 3.12a). Therefore, relationship between both sources seems difficult to be defined with the present available information. However, geochronological and geochemical data obtained from drill-core material in the western edge of the Río de la Plata craton show Paleoproterozoic igneous rocks related to an arc system that correspond to the Transamazonian orogenic cycle (Rapela et al. 2007). Therefore, we can assume that subduction related to the latter event may account for the chemistry of those LTi basalts.

References

Ardolino A, Mendía J (1989) Geología del área de San Ignacio y alredededores, Provincia de Misiones. Dirección Nacional de Geología y Minería (República Argentina), Unpublished Report, 19 p, Buenos Aires

Ardolino A, Miranda F (2008) Las cataratas del Iguazú: El agua grande. In: Sitios de Interés Geológico de la República Argentina, CSIGA (ed), Instituto de Geología y Recursos Minerales. Servicio Geológico Argentino, Buenos Aires, Anales 46(I), pp 377–389

Avila FM, Crivello JF, Portaneri JG (2008) Las minas de Wanda–Libertad: Piedras preciosas en Misiones. In: Sitios de Interés Geológico de la República Argentina, CSIGA (ed), Instituto de Geología y Recursos Minerales, Servicio Geológico Argentino, Buenos Aires, Anales 46(I), pp 391–399

Barberi F, Bizouard H, Varet J (1971) Nature of the clinopyroxene and iron enrichment in alkalic and transitional basaltic magmas. Contrib Miner Petrol 33: 93–107

Bellieni G, Piccirillo EM, Zanettin B (1981) Classification and nomenclature of basalts. IUGS, subcommission of the systematics of igneous rocks, circular 34. Contrib Miner Petrol 87:1–19

Bellieni G, Comin-Chiaramonti P, Marques LS, Melfi AJ, Piccirillo EM, Nardy AJ, Roisemberg A (1984) b) High- and low-TiO$_2$ flood basalts from the Paraná plateau (Brasil): petrology and geochemical aspects bearing on their mantle origin. Neues Jahrbuch für Mineralogie Abhandlungen 150:273–306

Bellieni G, Comin-Chiaramonti P, Marques LS, Martínez LA, Melfi AJ, Nardy AJR, Piccirillo EM, Stolfa D (1986) Continental flood basalt from the central-western regions of the Paraná plateau (Paraguay and Argentina): petrology and petrogenetic aspects. Neues Jahrbuch für Mineralogie Abhandlungen 154:111–139

Bossi J, Gaucher C (2004) The Cuchilla Dionisio, Uruguay: an allochtonous block accreted in the Cambrian to SW-Gondwana. Gondwana Res 7(3):661–674

Bossi J, Schipilov A (2007) Rocas ígneas básicas del Uruguay. Universidad de la República, Facultad de Agronomía, 364 p, Montevideo

Boynton WV (1984) Cosmochemistry of the rare earth elements: meteorite studies. In: Henderson P (ed) rare earth element geochemistry. Elsevier, Amsterdam, pp 63–114

Busby-Spera CJ, White JDL (1987) Variation in peperite textures associated with differing host-sediment properties. Bull Volc 49:765–775

Carlson RW, Esperança S, Svisero DP (1996) Chemical and Os isotopic study of Cretaceous potassic rocks from Southern Brazil. Contrib Miner Petrol 125:393–405

Carlson RW, Araujo ALN, Junqueira-Brod TC, Gasparf JC, Brod JA, Petrinovic IA, Hollanda MH, Pimentel MM, Sichel S (2007) Chemical and Isotopic relationships between peridotite xenoliths and magic-ultrapotassic rocks from Southern Brazil. Chem Geol 242:415–434

Casquet C, Rapela CW, Pankhurst RJ, Baldo EG, Galindo C, Fanning CM, Dahlquist JA, Saavedra J (2012) A history of Proterozoic s in southern South America: from Rodinia to Gondwana. Geosci Front 3(2):137–145

Comin-Chiaramonti P, Gomes CB (2005) Mesozoic to Cenozoic alkaline magmatism in the Brazilian Platform. Edusp/Fapesp, 750 p, San Pablo

Comin-Chiaramonti P, Cundari A, Piccirillo EM, Gomes CB, Castorina F, Censi P, De Min A, Marzoli A, Speziale S, Velázquez VF (1997) Potassic and sodic igneous rocks from Eastern Paraguay: their origin from the lithospheric mantle and genetic relationships with the associated Paraná flood tholeiites. J Petrol 38:495–528

Comin-Chiaramonti P, Marzoli A, Gomes CB, Milan A, Riccomini C, Velásquez VF, Mantovani MSM, Renne P, Tassinari CCG, Vasconcelos PM (2007) Origin of Post-Paleozoic magmatism in Eastern Paraguay. In: Foulguer GR, Jurdy DM (eds) Plates, plumes, and planetary processes. Geological society of america special papers 430. Boulder, Colorado, pp 603–633

Comin-Chiaramonti P, De Min A, Cundari A, Girardi VAV, Ernesto M, Gomes CB, Riccomini (2013) Magmatism in the Asunción-Sapucai-Villarrica Graben (Eastern Paraguay) Revisited: petrological, geophysical, geochemical and geodynamic inferences. Hindawi Publishing Corporation. J Geol Res 2013, Article ID 590835, 22 pages. http://dx.doi.org/10.1155/2013/590835

Courtillot V, Davaille A, Besse J, Stock J (2003) Three distinct types of hotspots in the Earth's mantle. Earth Planet Sci Lett 205:295–308

Cox D (1986) Descriptive model of basaltic Cu. In: Cox D, Singer D (eds) Mineral deposit models. U.S. Geol Surv, Bull 1693: 130

De la Roche H, Leterrier P, Grandclaude P, Marchal M (1980) A classification of volcanic and plutonic rocks using R1-R2 diagram and major element analysis. Its relationships with current nomenclature. Chem Geol 29:183–210

Duarte LC, Hartmann LA, Vasconcellos MAZ, Medeiros JTN, Theye T (2009) Epigenetic formation of amethyst-bearing geodes from Los Catalanes gemological district, Artigas, Uruguay, southern Paraná Magmatic Province. J Volcanol Geoth Res 184:427–436

Duarte LC, Hartmann LA, Medeiros JTN, Juchem PL (2014) Hydrothermal-epigenetic origin of amethyst and agate geodes in the Paraná volcanic province. In: Hartmann LA, Baggio SB (eds) Metallogeny and mineral exploration in the Serra Geral Group, IGEO/UFRGS: 303–320, Porto Alegre

Ernesto M, Pacca IG (1988) Paleomagnetism of the Paraná basin flood volcanics, southern Brazil. In: Piccirillo EM, Melfi AJ (eds) The Mesozoic flood volcanism of the Paraná basin: petrogenetic and geophysical aspects. Universidade de São Paulo, São Paulo, 229–255

Ernesto M, Raposo MIB, Marques LS, Renne PR, Diogo LA, de Min A (1999) Paleomagnetism, geochemistry and $^{40}Ar/^{39}Ar$ dating of the North-Eastern Paraná Magmatic Province: tectonic implications. 340

Ernesto M, Marques LS, Piccirillo EM, Molina EC, Ussami N, Comin-Chiaramonti P, Bellieni G (2002) Paraná Magmatic Province-Tristan da Cunha plume system: fixed versus mobile plume, petrogenetic considerations and alternative heat sources. J Volcanol Geoth Res 118:15–36

Escayola MP, Pimentel MM, Armstrong R (2007) Neoproterozoic backarc basin: Sensitive high-resolution ion microprobe U-Pb and Sm-Nd isotopic evidence from the Eastern Pampean Ranges, Argentina. Geology 35(6):495–498

Ferreira C, Dantas E, Pimentel M, Buhn B, Ruiz A (2008) Nd isotopic signature and U-Pb LA-ICPMS ages of Cambrian intrusive granites in the boundary between Brasília belt and Paraguay belt. 6° South American symposium on isotope geology. CD-Rom

Frey FA, Green DH, Roy SD (1978) Integrated models of basalt petrogenesis: a study of quartz tholeiites to olivine melilites from South Eastern Australia utilizing geochemical and experimental petrological data. J Petrol 19:463–513

Gastal MP, Lafon JM, Hartmann LA, Koester E (2005) Sm-Nd isotopic investigation of Neoproterozoic and Cretaceous igneous rocks from southern Brazil: a study of magmatic processes. Lithos 82:345–377

Gaucher C, Frei R, Chemale F Jr, Frei D, Bossi J, Martínez G, Chiglino L, Cernuschi F (2011) Mesoproterozoic evolution of the Río de la Plata craton in Uruguay: at the heart of Rodinia? Int J Earth Sci 100:273–288

Gentili C, Rimoldi H (1980) Mesopotamia. In: Turner JCM (ed) 2° Simposio de Geología Regional Argentina 1. Academia Nacional de Ciencias, Córdoba, pp 185–223

Ghiorso MS, Sack RO (1995) Chemical mass transfer in magmatic processes, IV, a revised and internally consistent thermodynamic model for the interpolation and extrapolation of liquid-solid equilibria in magmatic systems at elevated temperatures and pressures. Contrib Miner Petrol 119:197–212

Gibson SA, Thompson RN, Leonardos OH, Dickin AP, Mitchell JG (1995) The Late Cretaceous impact of the Trinidade mantle plume: evidence from large-volume, mafic, potassic magmatism in SE Brasil. J Petrol 36:189–229

Gibson SA, Thompson RN, Dickin AP, Leonardos OH (1996) Erratum to High-Ti and low-Ti mafic potassic magmas: key to plume-litosphere interactions and continental flood-basalt genesis. Earth Planet Sci Lett 141: 325–341

Gibson SA, Thompson RN, Day JA (2006) Timescales and mechanism of plume lithosphere interactions: $^{40}Ar/^{39}Ar$ geochronology and geochemistry of alkaline igneous rocks from the Paraná-Etendeka large igneous province. Earth Planet Sci Lett 251:1–17

Hanson GN (1978) The application of trace elements to the petrogenesis of igneous rocks of granitic composition. Earth Planet Sci Lett 38:26–43

Hartmann LA, da Cunha Duarte L, Massonne H-J, Michelin C, Rosenstengel LM, Bergmann M, Theye T, Pertille J, Arena K R, Duarte SK, Pinto VM, Barboza EG, Rosa M L, Wildner W (2010) Sequential opening and filling of cavities forming vesicles, amygdales and giant amethyst geodes in lavas from the southern Parana volcanic province, Brazil and Uruguay. Int Geol Rev 1–14

Hawkesworth CJ, Gallager K, Kelly S, Mantovani MSM, Peate D, Regelous M, Rogers N (1992) Parana magmatism and the opening of the South Atlantic and causes of Continental break-up. Geol Soc Spec Publ 68: 221–240. London

Herrmann C, Lagorio S, Segal S (2011). Mineralización de Cu en los basaltos de Serra Geral de la provincia de Corrientes, Argentina. Su caracterización en el contexto regional. 18° Congreso Geológico Argentino, Actas CD-Rom, Neuquén

Herrmann C, Lagorio S, Segal S, Cozzi, G (2013) Caracterización de la mineralización de cobre en los basaltos de la provincia de Corrientes. 10º Congreso Nacional de Geología Económica, Actas: 67–73, San Juan

Iacumin M, Piccirillo EM, Girardi VAV, Texeira W, Bellieni G, Echeveste H, Fernández R, Pinese JPP, Ribot A (2001) Early Proterozoic calc-alkaline and Middle Proterozoic tholeiitic dyke swarms from central-eastern Argentina: petrology, geochemistry, Sr-Nd isotopes and tectonic implications. Journal of Petrology 42:2109–2143

Iacumin M, De Min A, Piccirillo EM, Bellieni G (2003) Source mantle heterogeneity and its role in the genesis of Late Archean-Proterozoic (2.7-1.0 Ga) and Mesozoic (200 and 130 Ma) tholeiitic magmatism in the South American Platform. Earth Sci Rev 62:365–397

Iannizzotto NF, Rapela CW, Baldo EGA, Galindo C, Fanning CM, Pankhurst RJ (2013) The Sierra Norte-Ambargasta batholiths: late Ediacaran-Early Cambrian magmatism associated with Pampean transpressional tectonics. J S Am Earth Sci 42:127–143

Janasi VA, Freitas VA, Heaman LH (2011) The onset of flood basalt volcanism, Northern Parana basin, Brazil: U-Pb baddeleyite/zircon age for a Chapeco-type dacyte. Earth Planet Sci Lett 302:147–153

Jaques AL, Green DH (1980) Anhydrous melting of peridotite at 0–15 Kb pressure and the genesis of tholeiitic basalts. Contrib Miner Petrol 73:287–310

Kuiper KF, Deino A, Hilgen FJ, Krijgsman W, Renne PR, Wijbrans JB (2008) Synchronizing rock clocks of Earth history. Science 320:500–504

Lagorio SL (2008) Early Cretaceous alkaline volcanism of the Sierra Chica de Córdoba (Argentina): mineralogy, geochemistry and petrogenesis. J S Am Earth Sci 26:152–171

Lagorio SL, Leal PR (2005) Niveles peperíticos intercalados en los derrames lávicos de Serra Geral, provincia de Misiones. 16° Congreso Geológico Argentino, Actas 1: 847–850, La Plata

Lagorio SL, Vizán H (2011) El volcanismo de Serra Geral en la Provincia de Misiones: aspectos geoquímicos e interpretación de su génesis en el contexto de la Gran Provincia Ígnea Paraná-Etendeka-Angola. Su relación con el volcanismo alcalino de Córdoba (Argentina). Geoacta 36:27–53

Latorre C, Vattuone ME (1985) Apofilita, chabazita y minerales asociados de la Cantera Freyer, El Dorado, Misiones. Revista de la Asociación de Mineralogía, Petrología y Sedimentología 16 (1–2):17–25

Le Bas MJ, Le Maitre RW, Strekeisen A, Zanetin B (1986) A chemical classification of volcanic rock based on the total alkali-silica diagram. J Petrol 27:745–750

le Roex A, Cliff RA, Adair BJL (1990) Trsitan da Cunha, South Atlantic: geochemistry and petrogeneis of a basanite-phonolite lavas series. J Petrol 31:779–812

Lucassen F, Escayola MP, Romer RL, Viramonte JG, Koch K, Franz G (2002) Isotopic composition of Late Mesozoic basic and ultrabasic rocks from the Andes (23–32 S)—implications for the Andean mantle. Contrib Miner Petrol 143:336–349

Marengo, HG, Palma Y (2005) Diques y coladas en los basaltos de Serra Geral, área de San Ignacio, Misiones. 16° Congreso Geológico Argentino, Actas. Buenos Aires 1: 487–492

Marengo HG, Palma Y, Tchilinguirian P, Helms F, Kruck W, Roverano D (2005) Geología del área de San Ignacio, provincia de Misiones. 16° Congreso Geológico Argentino, Actas. Buenos Aires 1: 141–148

Marques LS, Dupre B, Piccirillo EM (1999) Mantle source compositions of the Parana Magmatic Province (southern Brazil): evidence from trace element and Sr-Nd-Pb isotope geochemistry. J Geodyn 28:439–458

Marzoli A, Melluso L, Morra V, Renne PR, Sgrosso I, D'Antonio M, Duarte Morais L, Morais EAA, Ricci G (1999) Geochronology and petrology of Cretaceous basaltic magmatism in the Kwanza basin (western Angola), and relationships with the Paraná-Etendeka continental flood basalt province. J Geodyn 28:341–356

McGregor ID (1974) The system $MgO-Al_2O_3-SiO_2$: solubility of Al_2O_3 in enstatite for spinel and garnet peridotite compositions. Am Mineral 59:110–119

Mena M, Orgeira MJ, Lagorio SL (2006) Paleomagnetism, rock-magnetim and geochemical aspects of early Cretaceous basalts of the Paraná Magmatic Province, Misiones, Argentina. Earth Planet Space 58:1283–1293

Peate DW (1997) The Paraná-Etendeka Province. In: Mahoney JJ, Coffin MF (eds) Large igneous provinces: continental oceanic and planetary flood volcanism. Geophys Monogr Am Geophys Union. Boulder, Colorado 100: 215–245

Peate DW, Hawkesworth CJ (1996) Lithospheric to asthenospheric transition in low-Ti flood basalts from Southern Paraná, Brazil. Chem Geol 127:1–24

Peate DW, Hawkesworth CJ, Mantovani MMS (1992) Chemical stratigraphy of the Paraná lavas (South America): classification of magma types and their spatial distribution. Bull Volc 55:119–139

Peate DW, Hawkesworth CJ, Mantovani MMS, Rogers NW, Turner SP (1999) Petrogenesis and stratigraphy of the high Ti/Y Urubici magma type in the Paraná flood basalt Province and implications for the nature of "Dupal"-type mantle in the South Atlantic Region. J Petrol 40:451–473

Piccirillo EM, Melfi AJ (1988) The Mesozoic flood volcanism from the Paraná basin: petrogenetic and geophysical aspects. Universidade de São Paulo, 600 p, São Paulo

Pinto VM, Hartmann LA (2014) Native copper mineralization in the Vista Alegre district, southernmost Brazil. In: Hartmann LA, Baggio SB (eds) Metallogeny and mineral exploration in the Serra Geral Group, IGEO/UFRGS, Porto Alegre, pp 355–368

Pinto VM, Hartmann LA, Wildner W (2011) Epigenetic hydrothermal origin of native copper and supergene enrichment in the Vista Alegre district, Paraná basaltic province, southernmost Brazil. Int Geol Rev 53:1163–1179

Rapela CW, Pankhurst RJ, Casquet C, Fanning CM, Baldo EG, González-Casado JM, Galindo C, Dahlquist J (2007) The Río de la Plata craton and the assembly of SW Gondwana. Earth-Scence Reviews 83:49–82

Rapela CW, Fanning CM, Casquet C, Pankhurst RJ, Spalletti L, Poiré D, Baldo EG (2011) The Río de la Plata craton and the adjoining Pan-African/brasiliano shield: their origins and incorporation into south-west Gondwana. Gondwana Res 20(4):673–690

Remesal MB, Wildner W, Chavez SB, Ardolino AA (2011) Provincia Magmática Paraná-Etendeka: Nueva propuesta estratigráfica. Misiones. Argentina. 18° Congreso Geológico Argentino, Actas CD-Rom, Neuquén

Renne PR, Ernesto M, Pacca IG, Coe RS, Glen JM, Prévot M, Perrin M (1992) The age of Paraná flood volcanism, rifting of Gondwanaland, and Jurassic-Cretaceous boundary. Science 258:975–979

Renne PR, Deckart K, Ernesto M, Feraud G, Piccirillo EM (1996a) Age of the Ponta Grossa dike swarm (Brazil) and implications to Paraná flood volcanism. Earth Planet Sci Lett 144:199–211

Renne PR, Glen JMS, Milner C, Duncan AR (1996b) Age of Edenteka flood volcanism and associated intrusions in southwestern Africa. Geology 24:659–662

Ringwood AE (1966) The chemical composition and origin of the Earth. In: Hurley PM (ed) Advances in earth sciences: contribution to the international conference on the earth sciences. Cambridge MIT Press, Cambridge, pp 287–356

Rocha-Júnior ERV, Puchtel IS, Marques LS, Walker RJ, Machado FB, Nardy AJR, Babinski M, Figueiredo AMG (2012) Re-Os isotope and highly siderophile element systematics of the Parana Continental Flood Basalts (Brazil). Earth Planet Sci Lett 337–338:164–173

Rocha-Júnior ERV, Marques LS, Babinski M, Nardy AJR, Figueiredo AMG, Machado FB (2013) Sr-Nd-Pb isotopic constraints on the nature of the mantle sources involved in the genesis of the high-Ti tholeiites from northern Parana Continental Flood Basalts (Brazil). J S Am Earth Sci 46:9–25

Stewart K, Turner SP, Kelley S, Hawkesworth CJ, Kirstein L, Mantovani MSM (1996) 3-D, ^{40}Ar-^{39}Ar geochronology in the Paraná continental flood basalt province. Earth Planet Sci Lett 143:95–109

Stormer JC, Nicholls J (1978) XLFRAC: A program for the interactive testing of magmatic testing of magmatic differentiation models. Computers and Geosciences 4:143–159

Sun SS, McDonough WF (1989) Chemical and isotopic systematics of oceanic basalts: implications for mantle composition and processes. In: Saunders AD, Norry MJ (eds) Magmatism in the Ocean Basins. Special Publication Geological Society of London 42: 313–345, London

Takahashi E, Kushiro I (1983) Melting of a dry peridotite at high pressure and basalt magma genesis. Am Min 68:859–879

Teruggi MM (1955) Los basaltos tholeiíticos de la provincia de Misiones. Notas del Museo de La Plata, Geología 18(70): 259–278

Thiede DS, Vasconcelos PM (2010) Paraná flood basalts: rapid extrusion hypothesis confirmed by new ^{40}Ar/^{39}Ar results. Geology 38(8):747–750

Thover E, Trindade RIF, Solum JG, Hall CM, Riccomini C, Nogueira AC (2010) Closing the Clymene ocean and bending a Brasiliano belt: evidence for the Cambrian formation of Gondwana, southeast Amazon craton. Geology 38(3):267–270

Turner SP, Regelous M, Kelley S, Hawkesworth CJ, Mantovani MSM (1994) Magmatism and continental break-up in the South Atlantic: high precision ^{40}Ar-^{39}Ar geochronology. Earth Planet Sci Lett 121:333–348

Turner SP, Peate DW, Hawkesworth CJ, Mantovani MSM (1999a) Chemical stratigraphy of the Paraná basalt succession in western Uruguay: further evicence for the diachronous nature of the Paraná magma types. J Geodyn 28:459–469

Turner SP, Kirstein L, Hawkesworth CJ, Peate DW, Hallinan S, Mantovani MSM (1999b) Petrogenesis of an 800 m lava sequence in eastern Uruguay: insights into magma chamber processes beneath the Paraná flood basalt province. J Geodyn 28:471–487

White RS, McKenzie DP (1989) Magmatism at rift zones: the generation of volcanic continental margins and flood basalts. J Geophys Res 94:7685–7730

Wildner W, Santos JOS, Hartmann LA, McNaughton NJ (2006) Clímax final do vulcanismo Serra Geral em 135 Ma: primeiras idades U-Pb em zircão. 43º Congresso Brasileiro Geologia, Extended Abstracts, Aracaju

Wildner W, Hartmann LA, da Cunha Lopes R (2007) Serra Geral magmatism in the Paraná Basin —a new stratigraphic proposal, chemical stratigraphy and geological structures. Prob West Gondwana Geol 1: 189–197. Gramado. Brazil

Chapter 4
Palaeoreconstruction of Pangea During the Early Cretaceous, and Location of Volcanism in Córdoba and Misiones Provinces with Respect to Seismic Structures in the Lower Mantle

Abstract To analyse the possible geodynamical processes that caused the volcanism in Misiones Province (north-eastern Argentina) involved in the context of Paraná Magmatic Province and the Early Cretaceous basalts of Córdoba Province (central Argentina), an absolute reconstruction of Pangea was performed on the basis of 7 high-quality palaeomagnetic poles of South America and Africa that belong to the localities with aged volcanic rocks about 130 Ma. The rest of the continents that formed Pangea were reconstructed on the basis of reliable parameters that have been published by other authors. It was considered that Africa had not significant longitudinal movements from at least the Early Cretaceous since there was not any margin of subduction during the history of this continent which slabs could have determined a large convective cell causing a longitudinal drift of this plate. The palaeogeographic reconstruction was analysed in the framework of a well-known model of seismic anomalies in the lowermost mantle. Cratons and other continental domains older than the Late Palaeozoic were also considered, taking into account that the Early Cretaceous volcanism in South America and Africa could have been also conditioned by the old lithospheric architecture of these continental plates.

Keywords Early cretaceous · Absolute palaeoreconstruction · Pangea · Volcanism · Lowermost mantle · Córdoba · Misiones · Argentina

To analyse the possible geodynamical processes that caused the PMP or Paraná–Etendeka–Angola large igneous province volcanism and the Early Cretaceous basalts of Córdoba, Vizán and Lagorio (2011) considered the geographical distribution of Precambrian and Early Palaeozoic units in South American plate, taking into account that the location of the volcanic rocks of the LIPs can be conditioned by the old lithospheric architecture (Favela and Anderson 2000). These units were located in South America according to Iacumin et al. (2003) and Rapela et al. (2007). An absolute reconstruction of Gondwana was performed for about 130 Ma based on palaeomagnetic data and considering that Africa did not experience

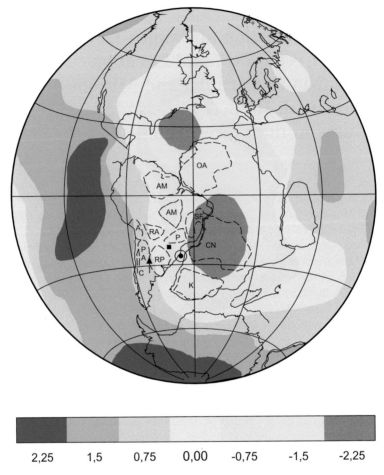

2,25   1,5   0,75   0,00   -0,75   -1,5   -2,25

Fig. 4.1 Absolute Palaeoreconstruction of Pangea for the 130 Ma. Crustal units older than Late Palaeozoic, in the geographic coordinates that would have had at 130 Ma, are also represented (*AM* Amazonia, *RP* Río de La Plata, *SF* San Francisco, *P* Paraná, *PA* Pampia, *C* Cuyania, *A* Arequipa, *RA* Rio Apa, *CN* Congo, *K* Kalahari, *EAC* East African craton). The solid dot within a circle shows the centre of eruption of the large igneous province Paraná–Etendeka–Angola. The square indicates the relative location of the volcanism in Misiones Province and the triangle the volcanism of the Sierra Chica of Córdoba. In a range from orange to blue colours a model of S seismic wave velocities (Masters et al. 1996) for the lower mantle is shown. Based on Vizán and Lagorio (2011)

significant longitudinal movement (Burke and Torsvik 2004; Torsvik et al. 2008, 2012) (Fig. 4.1).

To prepare this palaeoreconstruction, Vizán and Lagorio (2011) previously performed a proper selection of PPs. The 7 PPs selected by these authors are listed in Table 4.1 (note that these PPs are not precisely the same selected by Torsvik et al. 2008, 2012). All PPs selected have quality factors higher than 3 (see Van der

Table 4.1 Selected palaeomagnetic poles for the reconstruction at about 130 Ma (Fig. 3.11)

Geological unit or locality	PP latitude (°S)	PP longitude (°E)	A$_{95}$ (°)	Authors
Sierra Chica (Córdoba, Arg.)	86	75.9	3.3	Geuna and Vizán (1998)
Posadas Fm (Misones, Arg.)	87.6	150.8	3.5	Mena et al. (2006)
Northern Paraná (Brazil)	83	71.4	2.4	Ernesto et al. (1990)
Central Paraná (Brazil)	84.1	64.4	2.3	Ernesto et al. (1990)
Southern Paraná (Brazil)	84	106.2	1.5	Ernesto et al. (1990)
Alkaline Province (Paraguay)	85.4	62.3	3.1	Ernesto et al. (1996)
Kaoko Lavas (Africa)	48.3	86.6	3.2	Gidskehaug et al. (1975)

PP palaeomagnetic pole, A$_{95}$ 95 % confidence level of each PP. Mean Western Gondwana PP (in geographic coordinates of South Africa), $N = 7$, Lat. = 50.5° S, Long. = 83.74° E, A$_{95}$ = 2.1°, $K = 796.1$. The South American poles were reconstructed using the parameters listed in Torsvik et al. (2008). The confidence interval A$_{95}$ and the precision parameter Kappa (K) belong to the statistics of Fisher (1953)

Voo 1993) and belong to the localities with volcanic rocks; therefore, they are not biased by the inclination flattening that could affect the palaeomagnetic data of sedimentary rocks (e.g. Tauxe and Kent 2004). The PP of the Serra Geral Formation in Misiones (Mena et al. 2006) was obtained using only those directions that pass a reversal test. South American poles were transferred to African coordinates by using the reconstruction parameters listed in Torsvik et al. (2008), and after that, a mean PP ($N = 7$) was obtained that allowed to estimate an Euler pole to perform an absolute reconstruction of Africa–South America for 130 Ma (Lat. = 0°, Long. = 173.7° E, rotation angle = 39.5°).

The calculated Euler pole is similar to that of Torsvik et al. (2008) for the same age, indicating that the results are not strongly dependent on the selection of PPs. We did not use PPs from other continents of Gondwana, North America, Europe or Asia, but instead performed a reconstruction of these continents to Africa using the Euler poles of Torsvik et al. (2008) for the reconstruction of Pangea shown in Fig. 4.1. This figure shows an absolute reconstruction of Pangea (latitudinal and longitudinal) together with a model of the seismic structure of the mantle at a depth of 2750 km (Masters et al. 1996). South American terranes or cratons older than the Late Palaeozoic are also shown in Fig. 4.1. The African cratons bordering the Atlantic margin were drawn according to GMAP program and database (Torsvik and Smethurst 1999).

References

Bossi J, Gaucher C (2004) The Cuchilla Dionisio, Uruguay: an allochtonous block accreted in the Cambrian to SW-Gondwana. Gondwana Res 7(3):661–674

Burke K, Torsvik TH (2004) Derivation of large igneous provinces of the past 200 million years from long-term heterogeneities in the deep mantle. Earth Planet Sci Lett 227:531–538

Ernesto M, Pacca IG, Hiodo F, Nardy JR (1990) Paleomagnetism of the Mesozoic Serra Geral Formation, southern Brazil. Phys Earth Planet Inter 64:153–175

Ernesto M, Comin-Chiaramonti P, Gomes CB, Castillo AMC, Velazquez JC (1996) Palaeomagnetic data from the Central Alkaline Province, eastern Paraguay. In: Comin-Chiaramonti P, Gomes CB (eds) Alkaline magmatism in Central-Eastern Paraguay. University of Sao Paulo, Sao Paulo, pp 85–102

Favela J, Anderson DL (2000) Extensional tectonics and global volcanism. In: Boschi E, Ekstrom G, Morelli A (eds) Problems in geophysics for the New Millennium. Bologna Editrice Compositori, pp 463–498

Fisher RA (1953) Dispersion on a sphere. Proc Royal Soc A217:295–305

Gaucher C, Frei R, Chemale F Jr, Frei D, Bossi J, Martínez G, Chiglino L, Cernuschi F (2011) Mesoproterozoic evolution of the Río de la Plata craton in Uruguay: at the heart of Rodinia? Int J Earth Sci 100:273–288

Geuna SE, Vizán H (1998) New Early Cretaceous palaeomagnetic pole from Córdoba Province (Argentina): Revision of previous studies and implications for the South American database. Geophys J Int 135:1085–1100

Gibson SA, Thompson RN, Dickin AP, Leonardos OH (1996) Erratum to "High-Ti and low-Ti mafic potassic magmas: Key to plume-litosphere interactions and continental flood-basalt genesis". Earth Planet Sci Lett 141:325–341

Gidskehaug A, Creer KM, Mitchell J (1975) Palaeomagnetism and K-Ar ages of the South-West African basalts and their bearing on the time of initial rifting of the South Atlantic Ocean. Geophys J Roy Astron Soc 42:1–20

Iacumin M, De Min A, Piccirillo EM, Bellieni G (2003) Source mantle heterogeneity and its role in the genesis of Late Archean-Proterozoic (2.7-1.0 Ga) and Mesozoic (200 and 130 Ma) tholeiitic magmatism in the South American Platform. Earth Sci Rev 62:365–397

Lagorio SL (2008) Early Cretaceous alkaline volcanism of the Sierra Chica de Córdoba (Argentina): Mineralogy, geochemistry and petrogenesis. J S Am Earth Sci 26:152–171

Lagorio SL, Vizán H (2011) El volcanismo de Serra Geral en la Provincia de Misiones: aspectos geoquímicos e interpretación de su génesis en el contexto de la Gran Provincia Ígnea Paraná-Etendeka-Angola. Su relación con el volcanismo alcalino de Córdoba (Argentina). Geoacta 36:27–53

Marzoli A, Melluso L, Morra V, Renne PR, Sgrosso I, D'Antonio M, Duarte Morais L, Morais EAA, Ricci G (1999) Geochronology and petrology of Cretaceous basaltic magmatism in the Kwanza basin (western Angola), and relationships with the Paraná-Etendeka continental flood basalt province. J Geodyn 28:341–356

Masters G, Johnson S, Laske G, Bolton H (1996) A shear-velocity model of the mantle. Philo Trans R Soc Lond A 354:1385–1411

Mena M, Orgeira MJ, Lagorio SL (2006) Paleomagnetism, rock-magnetim and geochemical aspects of early Cretaceous basalts of the Paraná Magmatic Province, Misiones, Argentina. Earth Planet Space 58:1283–1293

Piccirillo EM, Melfi AJ (1988) The mesozoic flood volcanism from the Paraná Basin (Brazil): petrogenetic and geophysical aspects. Universidad de São Paulo, 600 p, San Pablo

Rapela CW, Pankhurst RJ, Casquet C, Fanning CM, Baldo EG, González-Casado JM, Galindo C, Dahlquist J (2007) The Río de la Plata craton and the assembly of SW Gondwana. Earth Sci Rev 83:49–82

Tauxe L, Kent DV (2004) A simplified model for the geomagnetic field and the detection of shallow bias in paleomagnetic inclinations: was the ancient magnetic field dipolar? In: Channell JET, Kent DV, Lowrie W, Meert JG (eds) Timescales of the paleomagnetic field. Geophysical Monograph Series, vol 145. American Geophysical Union, Boulder, Colorado, pp 101–115

Torsvik TH, Smethurst MA (1999) Plate tectonic modelling: virtual reality with GMAP. Comput Geosci 25:395–402

Torsvik TH, Müller RD, Van der Voo R, Steinberger B, Gaina C (2008) Global plate motion frames: towards a unified model. Rev Geophys 46. doi:10.1029/20007RG000227

Torsvik TH, Van der Voo R, Preeden U, Mac Niocaill C, Steinberger B, Doubrovine PV, van Hinsbergen DJ, Domeier M, Gaina M, Tohver E, Meert J, McCausland PJL, Cocks RM (2012) Phanerozoic polar wander, palaeogeography and dynamics. Earth Sci Rev 114:325–368

Van der Voo R (1993) Paleomagnetism of the Atlantic, Tethys and Iapetus Oceans. Cambridge University Press, Cambridge, 411 p

Vizán H, Lagorio SL (2011) Modelo geodinámico de los prcesos que generaron el volcanismo cretácico de Córdoba (Argentina) y la gran Provncia Ígnea Paraná, incluyendo el origen y evolución de la "pluma" Tristán. Geoacta 36:55–75

Chapter 5
Geodynamical Setting for the Tholeiites of Misiones Province (North-eastern Argentina) in the Context of the PMP and the Alkaline Volcanism of Córdoba Province (Central Argentina)

Abstract Different processes could have triggered magmatism during the Early Cretaceous in central and north-eastern Argentina. Distinct thermal processes could have possibly acted together in Paraná Magmatic Province (PMP): (1) the track of upwelling (strong upward heat flow) of a wide circuit of convection induced by a subduction in the western margin that cause the rifting of Western Gondwana and (2) a large amount of heat energy insulated by Pangea causing swelling and fragmentation of weak lithospheric zones, with high percentages of melting typical of the tholeiitic magmas. This volcanism must have emerged through lithospheric fracturing mainly in suture areas between old cratons, while South America plate was beginning to be dragged westward by a large convective roll. By contrast, the low volume of alkaline volcanic rocks in Sierra Chica of Córdoba (SCC) is in agreement with low melting degrees, in a geodynamic framework supporting edge-driven convection: (1) the presence of a large thickness contrast between the Río de La Plata craton and the Pampia terrain, (2) an environment of pull-apart basins and (3) a low rate of latitudinal and longitudinal velocity of South America.

Keywords Mantle convection · Magmatism · Alkaline basalts · Tholeiitic basalts · Edge-driven convection

The reconstructed position for South America at about 130 Ma (Fig. 4.1) shows that Pangea was geographically above the present zone of low seismic wave velocities in the lower mantle, as already pointed out by Vizán and Lagorio (2011). This is consistent with the proposal of thermal blanketing caused by this supercontinent (Anderson 1982; Gurnis 1988; Coltice et al. 2007); at 130 Ma years, there should have been a very high thermal energy beneath the continental lithosphere assembled into a single supercontinent.

It should be noted that in this figure, the centre of the Paraná-Etendeka-Angola Large Igneous Province (PEA LIP) is almost at a site of rifting of Western Gondwana (between South America and Africa plates), in a sector of stretched crust (Fig. 4.1) as pointed out by White and McKenzie (1989) in his classic work.

© The Author(s) 2016
S.L. Lagorio et al., *Early Cretaceous Volcanism in Central and Eastern Argentina During Gondwana Break-Up*, SpringerBriefs in Earth System Sciences, DOI 10.1007/978-3-319-29593-0_5

This location indicates that PEA LIP was one of the centres of dispersion of Gondwana continents (see Dietz and Holden 1970).

Rocha-Júnior et al. (2012) considered that the magmatism of the Paraná Magmatic Province (PMP) was due to the combined effects of edge-driven convection (King and Anderson 1995; King and Anderson 1998) and a large-scale mantle melting caused by the aggregation of continents to form Pangea (Anderson 1982; Coltice et al. 2007). However, according to King (2004) in areas where Pangea could have caused an overwhelmingly thermal insulation, any edge-driven convection would disappear. On the other hand, the South Atlantic Basin opened at ca. 130 Ma (e.g. Husson et al. 2012) probably due to a drag of South American plate to the west that could be determined by a large cylindrical convection (Froidevaux and Nataf 1981; Husson et al. 2012, among others).

This fact could involve an upwelling of sublithospheric melted material which would be coherent with new Os isotopic data in the PMP.

Towards the final stage of the PMP, the outpouring of the Esmeralda variety, characterized by a significant depletion in incompatible trace elements (as shown in Fig. 3.7e), let us infer a mix with N-MORB, as invoked by Peate and Hawkesworth (1996), which would be consistent with the beginning of the break-up of Western Gondwana and the future opening of the South Atlantic Ocean.

According to this evidence, and following Vizán and Lagorio (2011), we propose a large cylindrical convection to explain the PMP magmatism, whose origin would be linked to the subduction of a slab in the western margin of Gondwana (e.g. Rabinowicz et al. 1980; Nataf et al. 1981; Froidevaux and Nataf 1981; Husson et al. 2012). The model of Husson et al. (2012), proposed to analyse the westward movement of South America to support the growing of the Andean belt, could be also applied to explain the conditions that existed in the mantle at 130 Ma. The upwelling branch of the large convective cell in the mantle could have been driven by the heat energy insulated by Pangea (perhaps higher under the African cratons during the Cretaceous times), and the downwelling branch could have been fed by a down-going Pacific slab.

One possible reason why the African plate has not drifted longitudinally is because it has never had a subducted slab in its margins, and therefore, a large-scale convective roll was never generated under this plate.

In Fig. 5.1, different thermal processes are considered to have possibly acted together at approximately 130 Ma, which would overlap in the Paraná–Etendeka–Angola LIP: (1) the track of upwelling (strong upward heat flow) of the wide circuit of convection induced by the subduction in the western margin that cause the rifting of Western Gondwana and (2) the heat energy insulated by Pangea (Fig. 4.1). This process could have uplifted the sublithospheric material as suggested by the Os isotopic data together with a swelling and fragmentation of weak zones of the lithosphere due to the thermal energy insulated by Pangea (see Anderson 1994). In those areas, the thermal energy was high, determining the high percentages of melting, typical of the tholeiitic magmas that conform the large igneous provinces. Its volcanism must have emerged through lithospheric fracturing mainly in suture

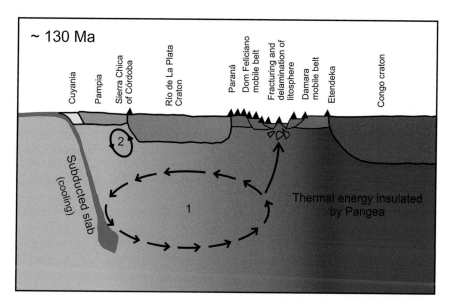

Fig. 5.1 Approximately east–west section of Western Gondwana (South America and Africa) at about 130 Ma with the location of crustal units covered by Phanerozoic deposits. Thickness of Río de la Plata craton is based on Favetto et al. (2008) and that of Congo–San Francisco craton on Schimmel et al. (2003). On the western margin of Gondwana, a subducted slab causing a lateral cooling of the asthenosphere; this, together with the thermal energy trapped by Pangea, would induce a large amplitude convective cell: (1) the upwelling branch of this cell would be in the centre of dispersion of Paraná–Etendeka–Angola large igneous province (PEALIP). The volcanism of this LIP is indicated by *black triangles*. In the Sierra Chica de Córdoba, the volcanism is only linked to edge-driven convection (2) due to the different thicknesses of Río de la Plata craton and Pampia terrane. Based on Vizán and Lagorio (2011)

areas between old thick cratons, because the South American plate was beginning to be dragged westward by a large convective roll (Fig. 5.1).

The volume of volcanic rocks in the SCC is much lower than that in the PMP; its alkaline nature indicates lower melting degrees than tholeiites from the LIP. This is consistent with the distance of SCC to the centre of higher thermal energy (Fig. 5.1) and its proximity to the western margin of Gondwana where a sinking subducted slab caused the cooling of the asthenosphere. This means that the magmatism of the SCC must have been caused by a different process from that of the PEA LIP.

To analyse the geodynamic processes involved in the genesis of the Lower Cretaceous basalts of the Córdoba province, it was considered that Favetto et al. (2008) using geological information and magneto-telluric data, interpret a sudden change in the observed resistivity in the SCC area, which would mark the boundary between the Río de la Plata craton and the Pampia terrane. The Río de la Plata craton is a resistive lithospheric block that would extend till 150 km in depth, while the Pampia terrane would be a heterogeneous structure with a high resistivity layer till 6 km and an underneath conductive layer till about 70 km. According to Vizán and Lagorio (2011), this strong contrast between the thicknesses of the mentioned blocks indicates that in the volcanic process of SCC an essentially edge-driven convection could have operated, in conditions of much lesser amount of heat than that characterizing the PMP volcanism (Fig. 5.1). However, other conditions are necessary for the development of this type of convection.

Edge-driven convection was considered by King and Anderson (1995) to explain volcanic processes that generated flood basalts in large areas or in relatively restricted depocenters, and this model has also been considered to explain volcanic events involving alkaline basalts (Missenard and Cadoux 2011). In addition to a contrast between lithospheric thicknesses for the development of this convection, at least two other conditions should exist: (1) the process should occur in a pull-apart tectonic setting (King and Anderson 1995) and (2) the movement of the plate where this convection is developed should be very slow (King and Anderson 1998). Edge-driven convection is induced by the temperature contrast at the vertical wall separating the cold craton from the warmer asthenosphere (King and Anderson 1998), and so it is a relatively weak instability and a fast relative motion between the lithosphere (craton and thin lithosphere) and the underlying asthenospheric mantle may produce a shear coupling that completely overwhelms this convection (King and Anderson 1998).

In the case of alkali basalts of Córdoba, its outpouring possibly occurred in pull-apart basins, according to Martino et al. (2014). Respect to the rate of speed of South America plate for the Early Cretaceous, it can be calculated, albeit with uncertainties, using an apparent polar wander path (APWP) in a frame of absolute drift (latitudinal and longitudinal movements of a plate).

To calculate the rate of speed of South America plate, it was considered the track extending from 200 to 60 Ma of the global APWP of Torsvik et al. (2012). Its PP's were referred to Africa since, as it was mentioned above, this continent has practically a negligible longitudinal drift. South America was reconstructed to present Africa geographical coordinates, using the parameters of the list of Torsvik et al. (2012), and the orientations and latitudinal locations of both continents were determined with the PP's of the APWP.

It is noteworthy that the used APWP was constructed averaging poles within 10 million years windows. Taken into account that according to Van der Voo (1993) using palaeomagnetism, it is possible to determine the drift of a plate with a reasonable resolution of approximately 20 million years, periods of this range were considered to analyse the absolute movements of South America (Fig. 5.2). Every versions of South American APWP show a change in motion between 140 and 120 Ma (e.g. Torsvik et al. 2012) defining a "hairpin" or "hook". This interval in the APWP was selected as one of the periods to analyse. To calculate the drift rate, the dam of Río Tercero (present geographic coordinates at Lat. = 32.2 °S and Lon. = 64.47 °W) was taken as a reference. Notice that the calculated rate is reduced practically to zero between 140 and 120 Ma (Fig. 5.2).

Thus, the outpouring of alkali basalts of Córdoba would have occurred in a geodynamic framework in which different conditions support a process of edge-driven convection during the Early Cretaceous: (1) the presence of a large

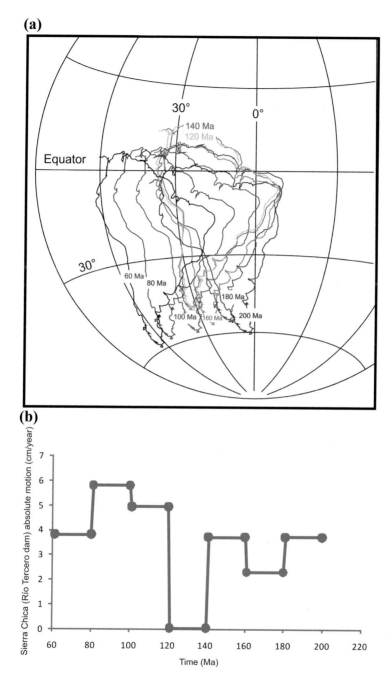

Fig. 5.2 a Absolute movements (latitudinal and longitudinal) of South America between 200 and 60 Ma. **b** Rate drift velocity referred to the Rio Tercero dam for a geological time between 200 and 60 Ma

thickness contrast between the Río de La Plata craton and the Pampia terrane, (2) an environment of pull-apart basins and (3) a low rate of latitudinal and longitudinal velocity of South America.

Therefore, we propose two distinct processes that triggered magmatism during the Early Cretaceous in different localities of Argentina. Both are consistent with Anderson (2001) idea that the convection in the asthenospheric mantle is organized by the above lithospheric plates. In other words, the patterns of convection in the sublithospheric mantle should be considered as the result and not the cause of plate tectonics.

References

Anderson DL (1982) Hotspots, polar wander, mesozoic convection and the Geoid. Nature 297:391–393

Anderson DL (1994) Superplumes or supercontinents? Geology 22:39–42

Anderson DL (2001) Top-down tectonics? Science 293:2016–2018

Coltice N, Phillips BR, Bertrand H, Ricard Y, Rey P (2007) Global warming of the mantle at the origin of flood basalts over supercontinents. Geology 35:391–394

Dietz R, Holden JC (1970) Reconstruction of Pangea: break-up and dispersion of continents, permian to present. J Geophys Res 75(26):4939–4956

Favetto A, Pomposiello C, López de Luchi MG, Booker J (2008) 2D Magnetotelluric interpretation of the crust electrical resistivity across the Pampean—Río de la Plata suture, in central Argentina. Tectonophysics 459:54–65

Froidevaux C, Nataf HC (1981) Continental Drift: what driving mechanism? Geol Rundsch 70:166–176

Gurnis M (1988) Large-scale mantle convection and the aggregation and dispersal of supercontinents. Nature 332:695–699

Husson L, Conrad CP, Faccenna C (2012) Plate motions, Andean orogeny and volcanism above the South Atlantic convection cell. Earth Planet Sci Lett 317–318:126–135

King SD (2004) Understanding the edge-driven convection hypotheses. www.MantlePlumes.org, last revision March, 2004

King SD, Anderson DL (1995) An alternative mechanism of flood basalt formation. Earth Planet Sci Lett 136:269–279

King SD, Anderson DL (1998) Edge-driven convection. Earth Planet Sci Lett 160:289–296

Martino RD, Guereschi AB, Carignano CA, Calegari R, Manoni R (2014) La estructura de las cuencas extensionales cretácicas de las Sierras de Córdoba. In: Martino RD, Guereschi AB (eds) Geología y Recursos Naturales de la Provincia de Córdoba, Relatorio del 19° Congreso Geológico Argentino: 513-538. Asociación Geológica Argentina, Córdoba

Masters G, Johnson S, Laske G, Bolton H (1996) A shear-velocity model of the mantle. Philos Transac Royal Soc Lond A 354:1385–1411

Missenard Y, Cadoux M (2011) Can Moroccan Atlas lithospheric thinning and volcanism be induced by edge-driven convection? Terra Nova. doi:10.1111/j.1365-3121.2011.01033.x

Nataf HC, Froidevaux C, Levrat JL, Rabinowicz M (1981) Laboratory convection experiments: Effect of lateral cooling and generation of instabilities in the horizontal boundary layers. J Geophys Res 86:6143–6154

Peate DW, Hawkesworth CJ (1996) Lithospheric to asthenospheric transition in low-Ti flood basalts from Southern Paraná, Brazil. Chem Geol 127:1–24

Rabinowicz M, Lago B, Froideveaux C (1980) Themal transfer between the continental asthenosphere and the oceanic subducting lithosphere: Its effect on subcontinental convection. J Geophys Res 85:1839–1853

Rocha-Júnior ERV, Puchtel IS, Marques LS, Walker RJ, Machado FB, Nardy AJR, Babinski M, Figueiredo AMG (2012) Re-Os isotope and highly siderophile element systematics of the parana continental flood basalts (Brazil). Earth Planet Sci Lett 337–338:164–173

Schimmel M, Asssumpçao M, Van Deccar JC (2003) Seismic velocity anomalies beneath SE Brazil from P and S wave travel time inversions. J Geophys Res 108(B4):2191. doi:10.1029/2001JB000187

Torsvik TH, Van der Voo R, Preeden U, Mac Niocaill C, Steinberger B, Doubrovine PV, van Hinsbergen DJ, Domeier M, Gaina M, Tohver E, Meert J, McCausland PJL, Cocks RM (2012) Phanerozoic polar wander, palaeogeography and dynamics. Earth Sci Rev 114:325–368

Van der Voo R (1993) Paleomagnetism of the Atlantic, Tethys and Iapetus Oceans, 411 pp. Cambridge University Press, Cambridge

Vizán H, Lagorio SL (2011) Modelo geodinámico de los prcesos que generaron el volcanismo cretácico de Córdoba (Argentina) y la gran Provncia Ígnea Paraná, incluyendo el origen y evolución de la "pluma" Tristán. Geoacta 36:55–75

White RS, McKenzie DP (1989) Magmatism at rift zones: the generation of volcanic continental margins and flood basalts. J Geophys Res 94:7685–7730

Chapter 6
Conclusions

The volcanic events here described represent the main exposed sites of Early Cretaceous volcanism in central and eastern Argentina; one of them, in Sierra Chica of Córdoba (SCC), is of alkaline type, whereas the other is located in Misiones Province (MS) and forms part of the Paraná Magmatic Province (PMP), of tholeiitic nature.

Volcanism of Sierra Chica of Córdoba occurs within the Central Rift System. Also to the south, in Levalle basin, a thick Early Cretaceous volcanic pile lies buried in the subsurface. Other localities where Early Cretaceous is exposed are Sierra de las Quijadas and Cerrillada de las Cabras, in San Luis Province, within the Western Rift System. In all the cases, lavas are interbedded with sedimentary deposits, under rift tectonics. In SCC lava flows are frequently associated with scoria fall, pyroclastic and phreatomagmatic breccias as well as volcaniclastic deposits, within a strombolian-type volcanism.

The low volume and discontinuous spatial distribution of the lavas along the rift, as well as the geochemical features without peralkaline oversaturated products, characterize Sierra Chica as a low-volcanicity rift.

Volcanic rocks of Sierra Chica of Córdoba belong to various groups of rocks according to the composition of major elements: (1) alkali basalt—trachyte suite, (2) transitional basalt—latibasalt suite, (3) basanite—phonolite suite and (4) ankaratrites.

The primary magmas of the different groups have different contents of Zr, Hf and LREE and distinct ratios (e.g. Th/La, La/Ta and La/Nb). In each suite, the evolution must have taken place at crustal level(s) from different parental magmas, mainly through fractional crystallization in an open system magma chamber. The latter involves local mixing with less evolved magmas, as in RTF model, along with some amount of crustal contamination.

Melting degree is around 5–7 % to obtain primary magmas of the major groups, from a hydrous enriched mantle source peridotite, supporting garnet and phlogopite as residual phases. The slight differences between the various primary magmas of

S.L. Lagorio et al., *Early Cretaceous Volcanism in Central and Eastern Argentina During Gondwana Break-Up*, SpringerBriefs in Earth System Sciences, DOI 10.1007/978-3-319-29593-0_6

the different groups essentially reflect a heterogeneous mantle source, therefore, of lithospheric nature.

A new ^{40}Ar/^{39}Ar dating was performed on sanidine phenocrysts of a trachyte collected at the Almafuerte locality of Sierra Chica of Córdoba. The obtained radiometric age of 129.6 ± 1.0 Ma indicates that SCC alkaline volcanism slightly postdates the tholeiitic lava flows of the Paraná Magmatic Province. This new dating seems to enlarge the number of posthumous alkaline localities peripheral to the PMP, which seem to prevail in the whole region.

The potassic rocks of Sierra Chica are of high Ti, as ultrapotassic ones from northern Brazil (Alto Paranaíba and Goiás localities), despite the fact that the latter are Late Cretaceous/Palaeogene in age. On the other side, SCC volcanic rocks diverge from those potassic rocks from eastern Paraguay as these are of low Ti and therefore have distinct multi-element patterns as well as isotopic features, though both of them are Early Cretaceous and partially coeval.

Even though the volcanism of Sierra Chica is related to the Pampean mobile belt, its geochemistry does not support a metasomatism due to fluids or melts derived from an ancient subduction. Small-scale heterogeneity in the rocks of the Sierra Chica would be related to metasomatic events that could involve volatile-rich small-volume melts from the asthenosphere. Consequently, the subduction inherent to the Pampean Orogeny must not have affected the source of the alkaline magmas of Sierra Chica de Córdoba.

Early Cretaceous volcanism in eastern Argentina is exposed in Misiones and Corrientes Provinces, whereas it lies under the surface of Entre Ríos Province and Chaco–Paraná basin (southern Paraná), corresponding to the central and souther regions of PMP. In the subsurface, floods reach about 150 km east of the Sierra Chica of Córdoba, so that no field relations between tholeiitic and alkaline lavas are exposed.

The samples from Misiones presented in this study are of high- and low-Ti varieties, which coincide with the geographic location of this locality that belongs to the central and southern Paraná Magmatic Province. In Corrientes Province, instead, low-Ti lavas prevail. To the south, in Corrientes Province, low-Ti lavas prevail instead. The studied rocks from MS indicate that evolution of HTi and LTi magmas must have occurred from two different parental magmas through fractional crystallization at low pressures, with low water content and low oxygen fugacity, involving crustal contamination only for the Gramado variety, in the same way as was pointed out for the whole PMP by other authors. (La/Yb)$_{cn}$ and (La/Nb)$_{PM}$ ratios are compatible with garnet in the source, taking into account that part of it must come from ancient remnant slabs in the mantle.

Misiones Province is in a zone between the Río de la Plata craton and the Paranapanema or Paraná block with the Paraguay belt located between them; therefore, ancient subduction processes must have taken place, mainly during Late Proterozoic–Early Cambrian. Furthermore, even older subduction processes relative to the Nico Pérez terrane could have also affected the source of the PMP basalts from Misiones. The presence of various cratonic blocks and ancient terranes characterizes the central and southern regions of Paraná, showing a remarkable contrast respect to the northern region. Undoubtedly, this may have conditioned the geochemical differences in the source of the PMP basalts from both regions.

In the context of the break-up of Pangea, we suggest different genesis for the volcanic events in the PMP and SCC during the Early Cretaceous. Several concomitant processes caused the volcanism of the PMP; on one side, we consider that Pangea insulated a great amount of thermal energy of radiogenic origin that melted the sublithospheric mantle, and on the other hand, we assumed the existence of a large-scale cylindrical convection whose origin requires a subducting slab in the western margin of Gondwana. This large cell would have dragged the future South American plate causing the outpouring of melts from the sublithospheric mantle in areas of crustal weakness (ancient sutures) where the source must have been affected by the material of Precambrian–Early Cambrian slabs. During the final stages of the magmatic process, the action of the upwelling limb of this possible large-scale cylindrical convection must have determined the outpouring of a depleted asthenospheric component, which seems to be recorded in southern PMP and finally in the N-MORB.

By contrast, the genesis of magmas of SCC, close to the Pacific margin of Gondwana, is considered to be related to edge-driven convection, probably determined by the abrupt change in the lithospheric thickness between the Río de la Plata craton and the Pampia terrane.

Index

A
Absolute paleoreconstruction, 124
Alkali basalts, 28, 30, 39, 61, 67, 75, 132
Alkaline basalts, 28, 61, 111, 115
Ankaratrites, 26, 32, 49, 59, 137
Argentina, 2, 74, 87, 88, 134
Argentine volcanism, 131

B
Basalts, 2, 11, 14, 28, 30, 33, 58, 59, 61, 63,
  67, 69, 75, 76, 78, 92, 95, 107, 109, 111,
  113, 114, 116, 132, 138
Basanites, 26, 31, 33, 49, 56, 59, 61

C
Córdoba, 2, 9, 11, 67, 68, 71, 72, 74, 75, 112,
  123, 131, 132, 137, 138
Corrientes, 87, 93, 106, 108
Cretaceous, 1, 2, 11, 12, 21, 23, 32, 72, 131,
  137

E
Early cretaceous, 2, 5, 10–12, 23, 24, 72,
  74–76, 78, 89, 91, 113, 123, 132, 134, 138,
  139
Edge-driven convection, 4, 130–132, 139

G
Geochemistry, 25, 46, 95, 138
Geodynamics, 27, 69, 74
Gondwana, 2, 70, 95, 123, 125, 130, 131, 139

H
High-Ti basalts, 95, 113

L
Large igneous province, 1, 123, 131
Lithospheric mantle, 4, 69, 75, 115, 134, 139
Low-Ti basalts, 95, 108, 109, 114

M
Magmatism, 130, 131, 134
Mantle convection, 5
Misiones, 87, 89, 91, 95, 106, 108, 111–113,
  124, 125, 137, 138

P
Pangea, 2, 5, 125, 130, 131, 139
Paraná, 1, 4, 11, 71, 72, 75, 76, 87, 89, 91, 95,
  106, 108, 112–115, 137, 138
Petrogenesis, 61

S
San Luis, 2, 9, 10, 76, 77, 137
Sierra Chica, 2, 9, 11, 12, 23, 24, 27, 46, 49,
  68, 71, 72, 74, 76, 114, 137, 138
South America, 1, 5, 32, 123, 129–134

T
Tholeiitic basalts, 89, 93
Transitional basalts, 31, 68, 70

V
Volcanism, 1, 2, 5, 14, 24, 74, 89, 91, 123,
  124, 139

© The Author(s) 2016
S.L. Lagorio et al., *Early Cretaceous Volcanism in Central and Eastern
Argentina During Gondwana Break-Up*, SpringerBriefs in Earth
System Sciences, DOI 10.1007/978-3-319-29593-0